IS-136 TDMA Technology, Economics, and Services

For a complete listing of the *Artech House Mobile Communications Library*,
turn to the back of this book.

IS-136 TDMA Technology, Economics, and Services

Lawrence Harte

Adrian Smith

Charles A. Jacobs

Artech House

Boston • London

Library of Congress Cataloging-in-Publication Data
Harte, Lawrence.
IS-136 TDMA technology, economics, and services / Lawrence Harte.
p. cm. — (Artech House mobile communications library)
Includes bibliographical references and index.
ISBN 0-89006-713-9 (alk. paper)
1. Time division multiple access. 2. Wireless communication systems—Standards. I. Series.
TK5103.486.H37 1998
621.382—dc21 98-33856
CIP

British Library Cataloguing in Publication Data
Harte, Lawrence
IS-136 TDMA technology, economics, and services
1. Mobile communication systems 2. Cellular radio
I. Title
621.3'84'56

ISBN 0-89006-713-9

Cover design by Lynda Fishbourne

685 Canton Street
Norwood, MA 02062

International Standard Book Number: 0-89006-713-9
Library of Congress Catalog Card Number: 98-33856

10 9 8 7 6 5 4 3 2 1

To Jacquelyn Gottlieb, the love of my life; my parents Virginia and Lawrence M. Harte; my children Lawrence William and Danielle Elizabeth; and all the rest of my loving family. *Lawrence*

To my fellow long-term inmates at the RTP Residence Inn who hammered out the draft specification for a digital control channel, the members of TR45.3 who created IS-136, and the people who designed and built the initial phone and infrastructure equipment. This effort has now clearly paid off with a successful, unique, and versatile wireless communications system. Also to my fiancée Debs for providing her support during what has been quite a hectic time! *Adrian*

I dedicate this book to my wife Pam and two sons Michael and Phillip. Their love and support give me the confidence and focus I need to be successful while remembering that life is about family. *Charles*

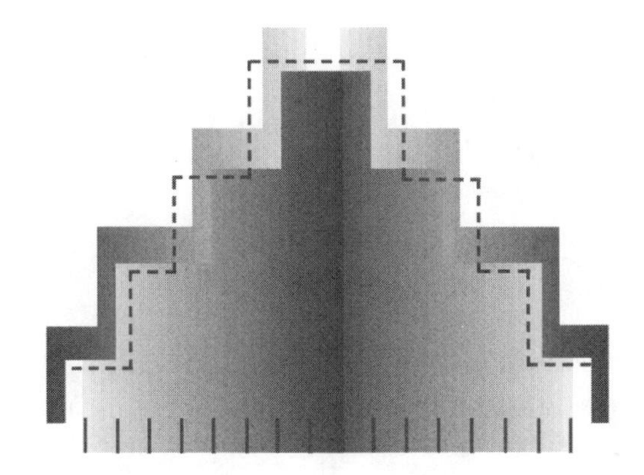

Contents

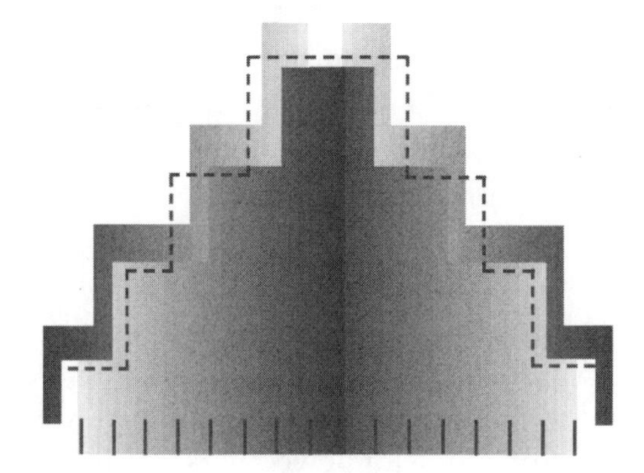

Preface

SINCE THE FIRST commercial *time division multiple access* (TDMA) telephones were introduced into the marketplace in 1991, the demand for TDMA digital telephones and service has grown by more than 60% per year. Predictions show that there will be over 300 million cellular and *personal communications service* (PCS) telephones in use worldwide by the year 2000; a large percentage of these will have digital transmission capability.

IS-136 TDMA technology has unique advantages and features that offer important choices for managers, technicians, and others involved with TDMA wireless telephones and systems. Accordingly, *IS-136 TDMA Technology, Economics, and Services* provides a detailed description of the IS-136 TDMA technology, demonstrates its economic benefits, and explains the new services that are, and will be, available through the application of this technology.

The book uses over 100 illustrations to explain IS-136 and IS-54 TDMA technology and its services. Furthermore, the book, whose technical content was reviewed by more than 100 industry experts, offers chapters that are carefully organized to help readers quickly find the detailed information they need. These chapters, which cover specific parts or applications of TDMA technology, may be read either consecutively or individually.

Chapter 1 introduces cellular and PCS technology. It covers industry terminology and the different types of analog and digital systems and discusses the factors that make IS-136 TDMA unique. It includes basic definitions of the major parts of analog cellular, digital cellular, and PCS systems. Chapter 1 is an excellent introduction for newcomers to cellular and TDMA technology.

Chapter 2 explains how analog cellular systems operate. It describes the radio channel structure, signaling messages, system parameters, and key features such as signaling messages, hand-off, and modulation.

Chapter 3 describes how IS-136 and IS-54 TDMA systems operate. It discusses these systems' radio channel structure, signaling messages, system parameters, and key features such as short messaging, private systems, and battery saving.

Chapter 4 discusses mobile telephone operation, design, and options. The descriptions cover the radio section, baseband signal processing, power supply, accessories and the call processing needed to make all the sections work together. Design options including *digital signal processor* (DSP) and *application-specific integrated circuit* (ASIC) tradeoffs are also covered.

Chapter 5 explains cellular networks, including their performance requirements and functional section descriptions. In addition, base stations and switching system equipment are described, and intranetwork signaling, internetwork connections, and key design and implementation options are discussed.

Chapter 6 provides an overview of the key testing requirements for mobile telephone, network equipment, and system field testing for TDMA systems.

Chapter 7 covers the data services supported by IS-136, including circuit-switched and packet-switched data.

Chapter 8 describes the requirements and applications for multimode, multiband telephones. This includes dual-band, dual-mode phones with special radio channel selection algorithms for intelligent roaming that finds IS-136 among other channels. These phones use AMPS, IS-136 cellular, and IS-136 PCS radio channels depending on what is available. Chapter 8 is a key part of IS-136 strengths in a multiband environment—giving exactly the same service where you get IS-136 TDMA—regardless of band.

Chapter 9 explains the marketing and economic factors that can significantly affect the rollout of TDMA technology. Included in this chapter's discussion are market demand estimates, mobile telephone costs, system equipment costs, operational costs, and key marketing considerations.

Chapter 10 identifies and analyzes the features and services that make TDMA such an attractive technology. These include private systems and the ways in which *digital control channel* (DCCH) features enable advanced services, including operation in the office environment.

Appendix A identifies the TDMA system standards; Appendix B lists the countries using AMPS and IS-136 TDMA; and Appendix C identifies control channel selection and camping criteria. Finally, Appendix D contains a description of reselection.

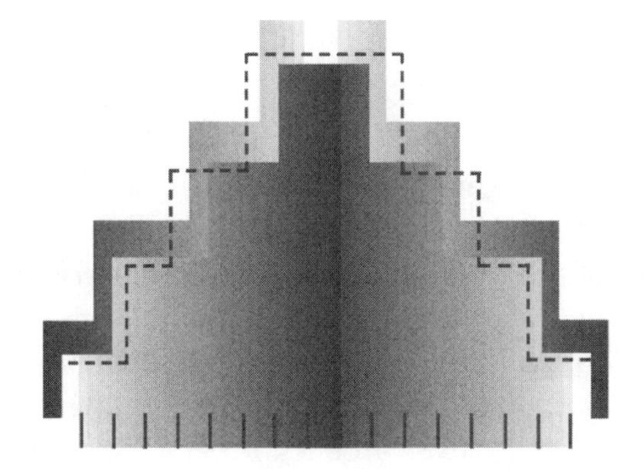

Acknowledgments

THE AUTHORS would like to thank Mike Raffel, Brian Daly, David Holmes, Glenn Blumstein, Michael Luna, John Myhre, and Michael Prise from AT&T Wireless Services for their help, advice, and support during the creation of this book. In addition, the authors extend a special thanks to Ian King for his insight in the testing chapter (Chapter 6) and to Larry Gillarde for his many contributions to this project. Angie Flom of the *North American Cellular Network* (NACN) and Chris Pearson of the *Universal Wireless Communications Consortium* (UWCC) also provided very useful information for which we are grateful.

We would also like to thank the following manufacturer representatives: Richard Levine of Beta Laboratories, Mike Cromie and Eric Stasik at Ericsson, Malcolm W. Oliphant of IFR, Jim Mullen of Hughes Network Systems, Raul Carr at Tellabs, Megan Matthews of Nokia Mobile Phones, and Karen Spitzner of Lucent Technologies.

In addition, we would like to express our gratitude to service provider representatives Ernesto Ramos of MovilNet and Richard Dreher at Evolving Systems.

Thanks also to the following research and consulting experts: Dave Crowe of Cellular Networking Perspectives, Elliott Hamilton at Strategis Group, Linda Gossack of the U.S. Department of Commerce, and Marty Nelson of Communication Services.

We would also like to thank the professionals at APDG, Inc.—Nancy Campbell, Lisa Gosselin, Judi Rourke-O'Briant, and Michael H. Sommer—for their assistance in the production of this book.

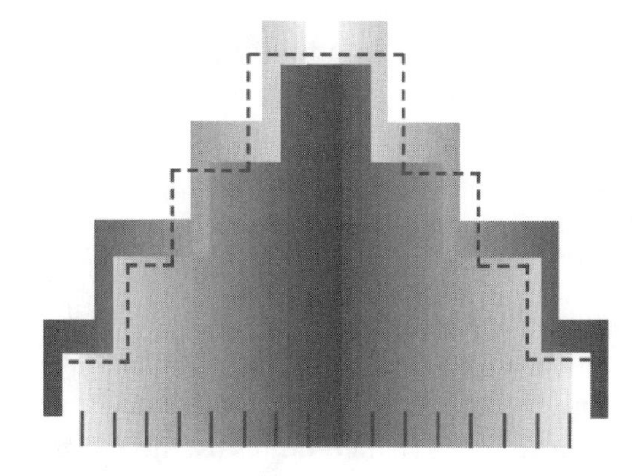

1

Introduction to Cellular and PCS

ALTHOUGH MOBILE RADIO has been in use for approximately 70 years and the cellular concept was conceived in the 1940s, public cellular mobile radio was not introduced in the United States until 1983. More modern electronic technology was needed to allow the introduction of the cellular and PCS systems we know today. Significant technology developments that are used in mobile radio technology come from broadcast radio, simplex two-way (one speaker at a time) and duplex (two-way simultaneous) transmission systems, radio system interconnection to the *public switched telephone network* (PSTN), radio channel selection through the use of trunking, and more efficient use of the radio frequency spectrum.

In the beginning of the twentieth century, mobile radio effectiveness was limited to shipboard use due to its high power requirements and large bulky tube radio technology. Automotive systems in the 1920s operated on 6V batteries with limited storage capacity. In addition, mechanical design of the tubes was susceptible to vibration that resulted in poor performance and reliability problems.

One of the first useful applications of mobile radio communications was implemented in 1928 by the Detroit police department. In this system, radio transmission was broadcast from a central location and could only be received by mobile radios. This effective system was a significant step in mobile radio communications. However, the beginning of the economic depression, prohibitive equipment power consumption requirements, and the need for vacuum tubes that could withstand vibration delayed the introduction of the first two-way mobile application until 1933 when the police department in Bayonne, New Jersey, introduced a simplex *amplitude modulation* (AM), push-to-talk system. Subsequently, the first *frequency modulation* (FM) mobile radio system was introduced in the Connecticut state police dept. at Hartford in 1940.

The first step toward mobile radio connection with the landline telephone network was established in St. Louis in 1946. This was called an "urban" system but only allowed three channels to be used at the same time.

Full-duplex (two-way simultaneous) mobile transmission did not occur until the mid 1950s when RCA completed a controversial contract for the City of Philadelphia police department. Although the system's technical specifications went beyond the technical capability at the time, a limited form of mobile FM modulated duplex communications was, nevertheless, developed.

Early radio systems' need for dedicated channels for a small group of users limited the efficiency of the radio spectrum. Radio channel trunking overcomes this limitation by affording the automatic sharing of several communications channels among a large group of users. During 1964, the *improved mobile telephone service* (IMTS) system was introduced. This system, which supported automatic channel selection and trunking and allowed full-duplex transmission, was the first real step toward mobile telephony as we know it today.

Nevertheless, the lack of available frequency bands restricted the total number of radios that could be supported on one system. Early mobile applications required much larger bandwidths than those needed today due to the lack of crystal-controlled frequency stability. Wide bandwidth requirements, up to 20 times larger, were required to allow for the frequency drift. Early on, the FCC believed that to serve the majority of users, much of the available commercial spectrum needed to be assigned to broadcast systems. Unfortunately, even though trunking and more stable transmitters increased the efficiency of spectrum utilization, a majority of the population could not be served with the early spectrum allocation. As a result, in 1976, people were placed on waiting lists to get a mobile phone.

Mobile radio had been in demand for many years, but its high equipment and use costs kept its market share small until the early 1980s. Market studies had shown a large demand at low prices; this demand has now been met by the low costs of cellular and PCS technology.

The key objectives of mobile radio systems have evolved to include new services and more efficient systems. There are several new types of cellular systems, and new frequency bands have become available that foster industry competition.

1.1 What is cellular?

A cellular system consists of three basic parts: mobile telephones, base stations, and a *mobile switching center* (MSC). Mobile telephones communicate by radio signals to nearby base stations. The base station converts these radio signals for transfer to an MSC via a wired or wireless communications link. The MSC coordinates and routes calls to other mobile telephones or to wired telephones that are connected to the PSTN. Figure 1.1 shows a basic cellular system.

The cellular concept allows small radio coverage areas to become part of a larger system by interconnecting them via the MSC. As a mobile phone moves from one radio coverage area to another, its radio frequency is changed from the frequency of the radio coverage area it is leaving to the frequency of the radio coverage area it is entering. This process is known as *hand-off*. Figure 1.2 shows the basic hand-off operation.

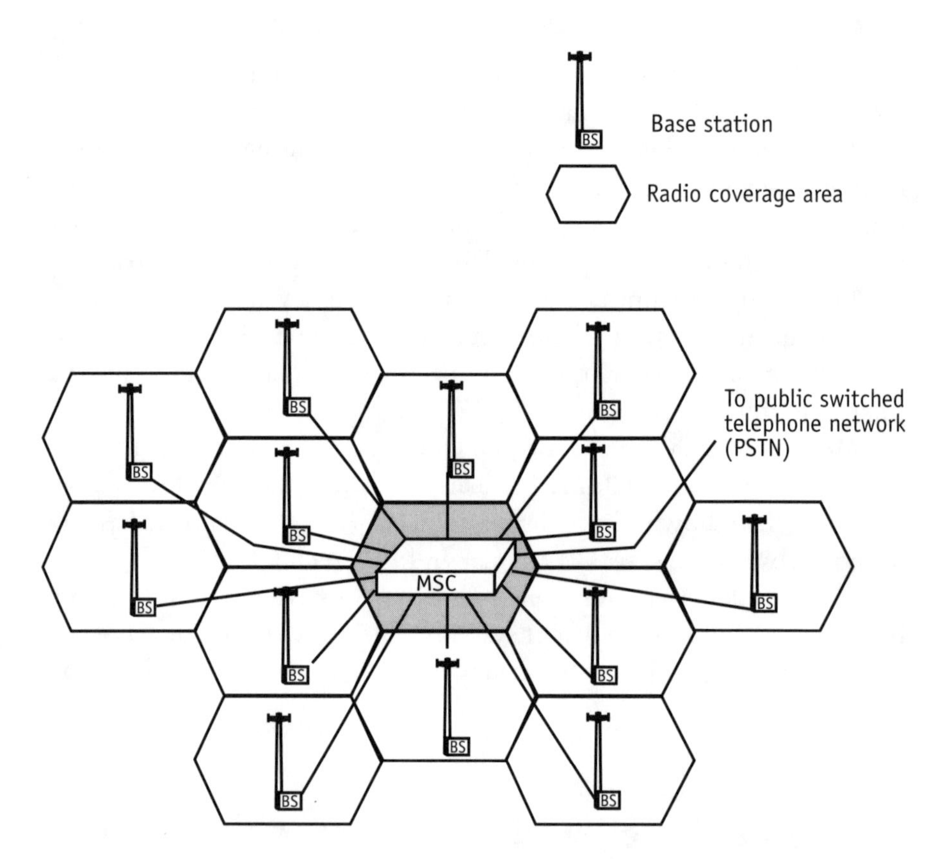

Figure 1.1 Basic cellular system.

1.2 What is PCS?

Personal communication services (PCS) are advanced services—such as longer battery life, enhanced voice quality, messaging (paging or wireless e-mail) services, and data services—that are not normally available on analog cellular systems. In 1993, the FCC defined a set of frequencies in the 1,900-MHz band (about double the frequency of cellular) that can be used to provide these types of services. There is nothing special about these PCS frequencies (as frequencies in the 1,900-MHz band are sometimes called), and with appropriate technology choice, exactly the same services can be offered in the 800-MHz *cellular* band for seamless PCS services throughout a wireless carrier's network.

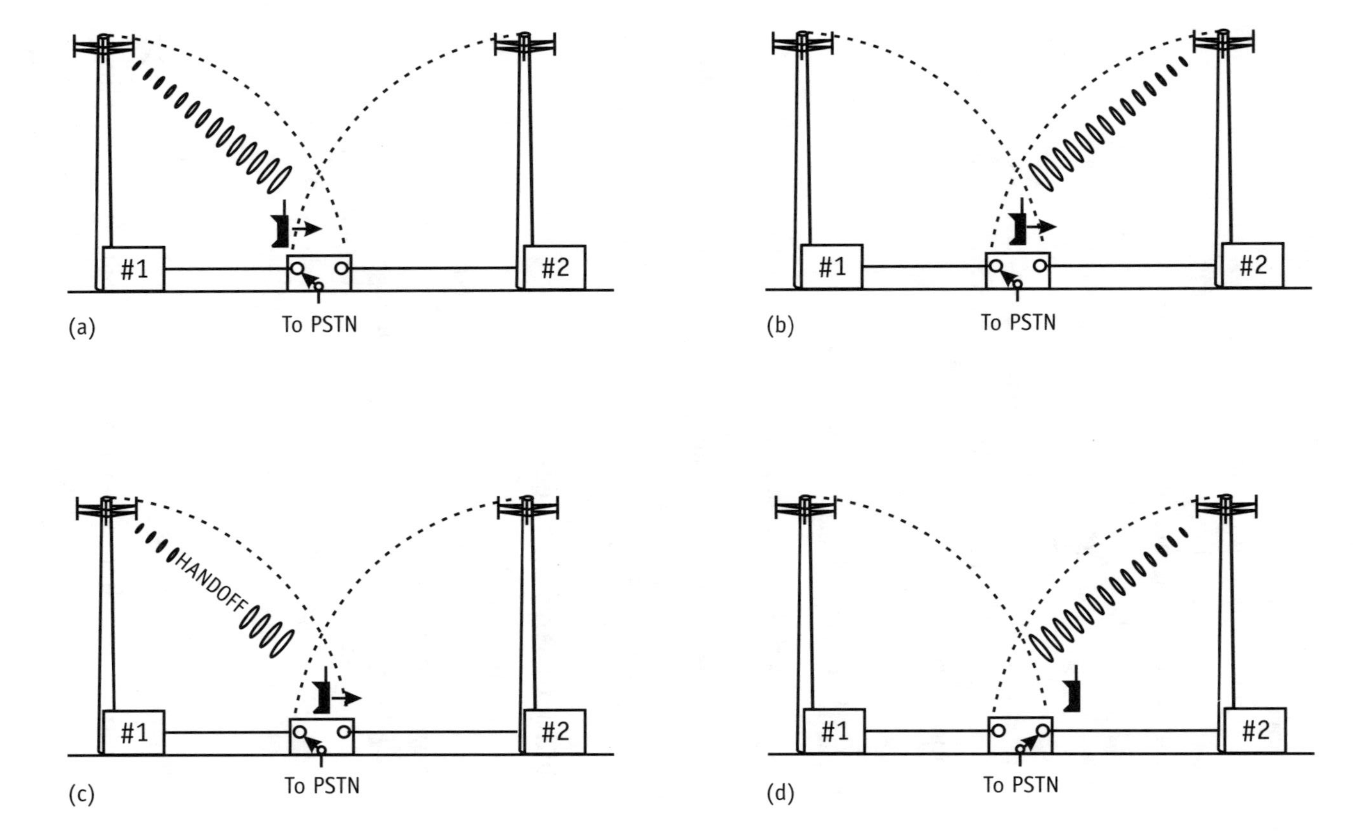

Figure 1.2 Cellular system hand-off: (a) base station #1 senses low power level; (b) base station #2 senses power level; (c) base station #1 sends hand-off message; (d) base station #2 begins transmission.

The 1,900-MHz frequency bands defined by the FCC have been divided into two categories of usage—licensed and unlicensed. Licensed frequencies guarantee the rights of the service operator to offer service without interference from other mobile telephones. Unlicensed bands allow any service provider the right to provide service with anticipated and reasonably coordinated interference from other mobile telephones.

Licenses for new 1,900-MHz radio channel frequency bands were auctioned by the federal government. Because there were existing licensed users in the radio spectrum chosen for service operators, the government requires the existing users to relocate over time. The existing users of this radio spectrum are primarily microwave point-to-point communications providers.

1.3 Cellular and PCS objectives

Many customers want to communicate wirelessly with other people by using voice and sending and receiving messages. To achieve the major objectives of the customer, cellular and PCS systems must be capable of simultaneously serving many customers in a cost-effective way with limited resources while providing advanced and simple-to-use services.

1.3.1 Ability to serve many customers

Mobile radio systems must have the ability to serve many customers with a limited number of radio channels. This involves coordinating the service requests of these customers and the sharing of radio channel frequencies. The cellular system coordinates mobile phones through the use of MSCs. The MSC contains information on the location and authorization of mobile phones to access the system. The MSC also coordinates the assignment of mobile phones to available radio channels at nearby base stations.

1.3.2 Efficient use of radio spectrum

Radio spectrum is a very limited resource. There are two ways a cellular system provides for the efficient use of radio spectrum: frequency reuse and access technology. The frequency reuse concept provides an increased system capacity through the continual reuse of radio frequencies. That is, the same frequency is used over and over again for different mobile phones in a system when they are geographically separated and the interference between them is at a minimum. To allow for more efficient use of the radio channels, advanced radio access technologies allow the simultaneous sharing of a single radio channel by multiple mobile phones. This allows the equipment dedicated for a single radio channel to serve several customers.

1.3.3 Nationwide compatibility

It is important for customers to have the ability to use their mobile phone throughout the United States and in various other countries. Nationwide compatibility is made possible through the use of industry standards and government regulation. The first industry standard was the analog cellular system *advanced mobile phone service* (AMPS). The government mandates that companies licensed to offer cellular service must use some of their radio channels to provide AMPS cellular service. As a result, AMPS cellular service is available throughout most of the United States and in over 50 other countries.

1.3.4 Widespread availability

Mobile telephones need to be available for purchase at many locations. The widespread availability of mobile phones has been made possible by competitive pricing, consumer acceptance, and simplified service activation. Since the commercial introduction of mobile phones in 1983, dozens of manufacturers have produced phones in a variety of cosmetic styles and at a low cost. This has resulted in the availability of different models of mobile phones through many specialty and retail locations. Preprogrammed mobile phones and over-the-air programming all assist in the sale of mobile phones in these locations.

1.3.5 Adaptability to high usage area requirements

The usage of mobile telephone service in particular regions varies based on a number of factors, including geography, environment, and time. For example, sparsely populated areas require less radio resources than urban centers, airports, office complexes, or busy freeways. To avoid congestion—which prevents a subscriber from making a call when all radios are being used—radio resources must be matched with expected call traffic. Thus, the highest call volume during "busy hour" must be anticipated.

1.3.6 Service to mobile vehicles and portable phones

The first mobile phones were high-power mobile telephones used mainly in vehicles. A higher output power mobile phone was required in early cellular systems since there were fewer cell sites spread across a wide geographical area. As traffic demands on cellular systems grew, companies have increased capacity by providing service with a greater number of smaller cell sites that can operate with lower power phones. Today's portable phones are able to use lower power levels to reduce interference, enhance battery life, and take advantage of smaller cell size.

1.3.7 Support of special services

The first mobile telephones were used primarily for voice (telephone) service. Modern wireless systems also provide advanced services, such as messaging and information services, using digital transmissions and intelligent network features.

1.3.8 Telephone quality of service

To achieve better consumer acceptance, mobile phones need to attain a quality level of service similar to that offered by wired (landline) telephone service. Analog AMPS mobile phones use FM radio, which provides a voice quality similar to landline telephone service when operated in good radio signal conditions. Unfortunately, radio signals do experience radio interference that sometimes reduces the quality of voice transmission. The new cellular and PCS systems use digital transmissions that provide the ability to detect, correct, and avoid radio distortions. This affords a more consistent level of voice quality.

1.3.9 Affordability

For cellular and PCS systems to be affordable to the average consumer, both the service cost and equipment cost must be reasonable. By utilizing more efficient technologies, the service usage cost of cellular and PCS systems has decreased over the past 10 years. In fact, due to the high-volume production of mobile phones, the cost of mobile phones has decreased by over 90% since 1983. As a result, mobile phones and service are affordable to a large number of consumers in the United States.

1.4 Forms of cellular and PCS

Cellular and PCS radio technologies offer various levels of voice, messaging, and data services. Some of these cellular systems use analog (direct modulation) transmission, while others use digital (digitized) communications. The basic types of systems used in the United States include AMPS, NAMPS, TDMA (IS-136), GSM, and CDMA (IS-95).

1.4.1 AMPS

The first commercial cellular technology in the United States was the analog (FM) AMPS system. In 1971, AT&T proposed a cellular radio telephony system to the FCC.[1] With minor changes, this system evolved to become the U.S. commercial cellular system, which launched service in Chicago in October 1983.

The AMPS system divides the frequency bandwidth into 30-kHz-wide radio channels. When in operation, each radio channel serves a single user. This division of radio channels is called *frequency division multiple access* (FDMA).

There are two types of channels in the AMPS system, dedicated control channels and FM-modulated voice channels. Dedicated control channels send paging messages and coordinate access to the system. The control channels use *frequency shift keying* (FSK) modulation that sends digital messages at a data rate of 10 Kbps. After the control channel has coordinated access, the mobile phone is assigned to a radio channel where it uses FM modulation to transfer voice signals. Figure 1.3 shows the basic radio system structure of an AMPS cellular system.

---

1. Dr. George Calhoun, *Digital Cellular Radio,* Norwood, MA: Artech House, 1988, pp. 50–51.

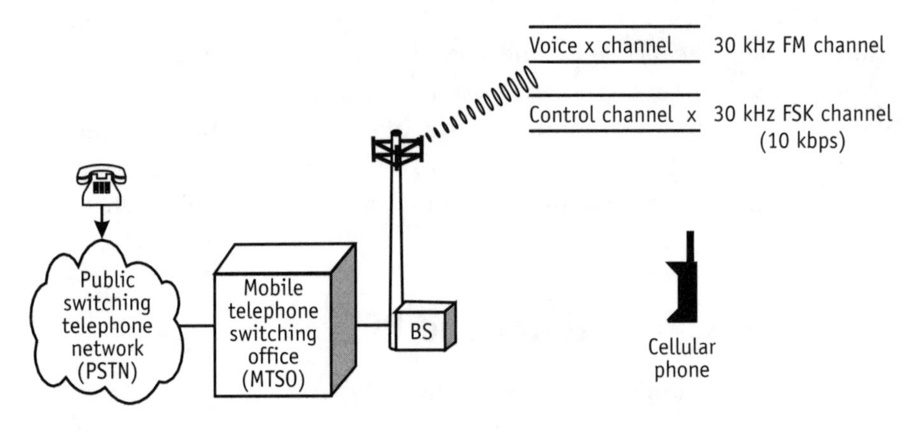

Figure 1.3 AMPS cellular system.

1.4.2 NAMPS

The IS-88 *narrowband AMPS* (NAMPS) standard combines AMPS and NAMPS functionality. NAMPS is an FDMA system commercially introduced by Motorola in late 1991. It is currently being deployed in parts of the United States and various countries worldwide. Like the existing AMPS technology, NAMPS uses analog FM radio for voice transmissions. The distinguishing feature of NAMPS is its use of a "narrow" 10 kHz bandwidth for radio channels, which are about one third the size of AMPS channels. More of these narrower radio channels can be installed in each cell site, allowing NAMPS systems to serve more mobile phones without adding cell sites. NAMPS also shifts some control commands to the subaudible frequency range to facilitate simultaneous transmissions.

In 1991, the first NAMPS standard, named IS-88, evolved from the EIA-553 AMPS specification. The IS-88 standard identified parameters needed to begin designing NAMPS radios, such as radio channel bandwidth, type of modulation, and message format. During development, the NAMPS specification benefited from the narrowband JTACS radio system specifications. During the following years, advanced features such as ESN authentication and caller ID were added to the NAMPS specification.

One of the IS-88 system's greatest advantages arises from its use of AMPS system FM technology. The employment of this existing technology minimized changes in mobile phones and base stations, where other standards using new technologies have required extensive changes. The new mobile phones have dual-mode capability and operate either on existing AMPS radio channels or new NAMPS voice channels. NAMPS systems use an analog control channel very similar to that of existing AMPS systems.

The IS-88 system has added a new *narrowband analog voice channel* (NAVC) radio channel. The 10-kHz-wide radio channel offers better sensitivity, noise rejection, and radio channel efficiency, and it can transfer simultaneous voice and data information. Figure 1.4 shows an overview of the NAMPS system.

1.4.3 TDMA

The IS-136 TDMA standard combines digital TDMA and AMPS technologies. In 1990, the first TDMA digital systems were introduced in North America. Today, over 36 countries use TDMA cellular technology.

IS-136 evolved from the TDMA standard IS-54. The IS-54 standard identified the critical radio parameters (e.g. time slot structure, type of radio channel, and modulation) and included features like calling number identity, message waiting indicator, authentication, voice privacy, and a 300-percent capacity improvement over analog cellular. In order to include further PCS features such as enhanced battery life, 800- and

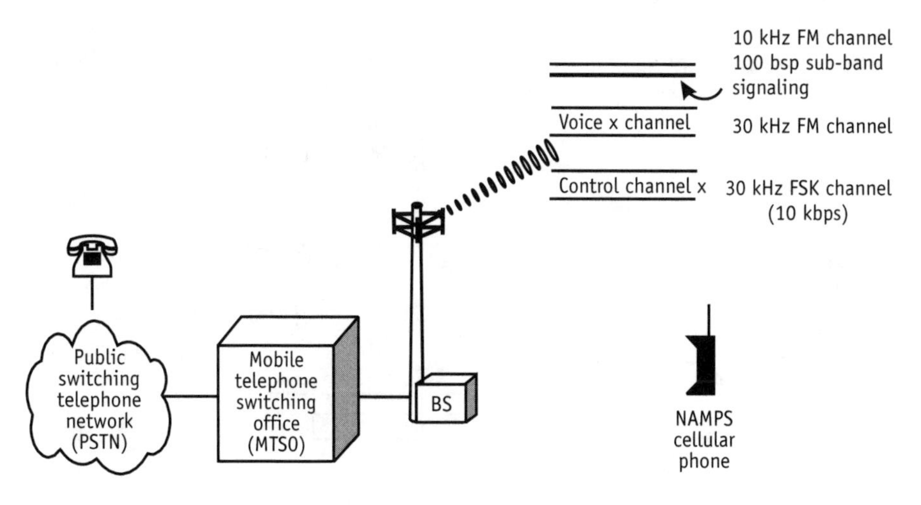

Figure 1.4 NAMPS (IS-88) radio system.

1,900-MHz operation, and private system support, the digital control channel and the IS-136 series were developed. Because of the evolution process, IS-136 phones can operate in an analog or digital environment in either the 800- or 1,900-MHz frequency band, while the same system infrastructure can maintain backward compatibility with the base of existing subscribers.

A primary feature of the IS-136 system is its coexistence with existing AMPS systems. IS-136 radio channels retain the same 30-kHz bandwidth as AMPS channels, and both AMPS and IS-136 service can be offered from the same system and cell sites to serve both new and existing customers. In addition, IS-136 phones are able to use both AMPS and IS-136 channels depending on available system resources providing a wide geographic dual-mode capability.

TDMA digital traffic channels are divided into frames with six time slots. Every communications channel consists of two 30-kHz wide channels, a forward channel (from the cell site to the mobile phone), and a reverse channel (from the mobile phone to the cell site). The time slots between forward and reverse channels are related so that the mobile phone does not simultaneously transmit and receive. Figure 1.5 presents an overview of the IS-136 TDMA system.

1.4.4 Global system for mobile communications

The *global system for mobile communications* (GSM) is a European digital cellular standard that uses a different form of TDMA than IS-136. While

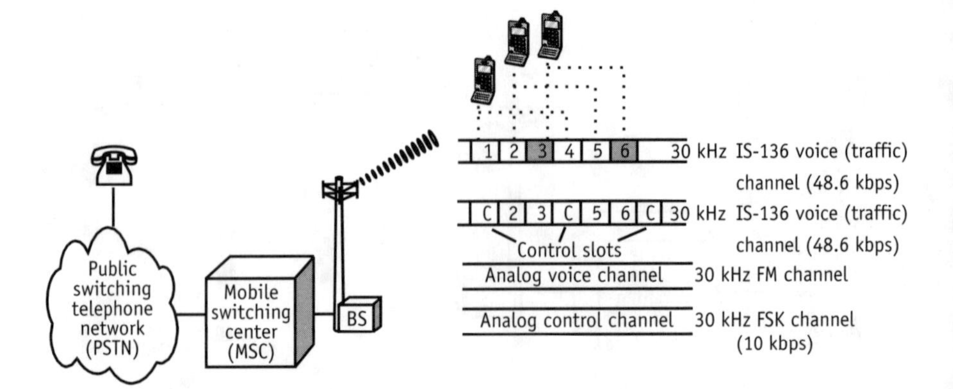

Figure 1.5 IS-136 TDMA cellular system.

GSM was initially developed for use in Europe, it is now available in over 109 countries throughout the world, including its use in PCS systems (PCS-1900).

A primary feature of the GSM system is its use of a single type of digital radio channel. Each 200-kHz-wide GSM digital radio channel is divided into frames with eight time slots. Every GSM channel consists of radio channels, a forward channel (from the cell site to the mobile phone), and a reverse channel (from the mobile phone to the cell site). The time slots between forward and reverse channels are related so that the mobile phone does not simultaneously transmit and receive.

The GSM system also uses portable *subscriber identity module* (SIM) cards that contain the identity of the customer. These cards can be removed from a GSM phone and used in another GSM mobile phone, thereby allowing a customer to use any GSM mobile phone for service. Figure 1.6 provides an overview of the GSM system.

1.4.5 Code division multiple access

The IS-95 *code division multiple access* (CDMA) system uses spread spectrum (wide RF bandwidth) digital cellular technology, which was commercially deployed in Korea in 1995 and in the United States in 1996. The IS-95 CDMA system combines a new digital CDMA radio channel and AMPS functionality. In 1998, CDMA systems were in use in over 10 countries.

The new CDMA digital RF channel provides both control and voice functionality. This spread spectrum radio channel is a wide 1.23-MHz

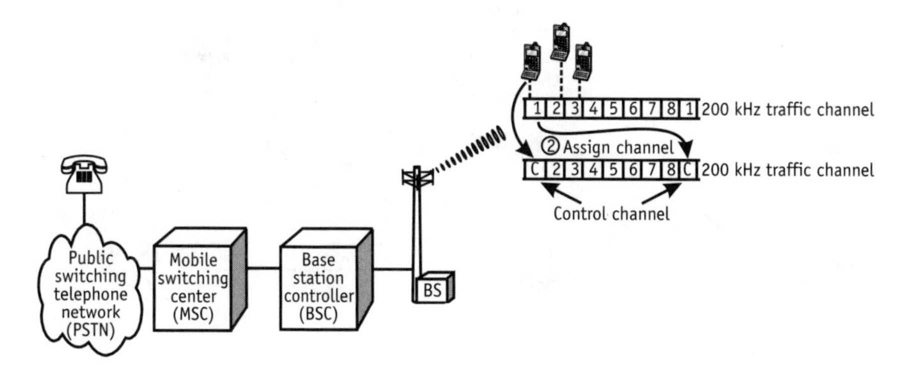

Figure 1.6 GSM cellular system.

digital radio channel. CDMA differs from other technologies in that it multiplies (spreads) each signal with a unique *pseudorandom noise* (PN) code that identifies each communications channel within the RF channel. This allows a CDMA radio channel to transmit digitized voice and control signals on the same frequency band. Each CDMA radio channel contains the signals of many ongoing calls (voice channels) together with pilot, synchronization, paging, and access (control) channels. CDMA mobile phones select the signal they are receiving by correlating the received signal with the proper PN sequence. This correlation process enhances the selected signal code and leaves other codes unenhanced. Figure 1.7 depicts an IS-95 CDMA cellular system.

1.5 Cellular and PCS parts

There are four basic parts to a cellular or PCS system: mobile phones, base stations, mobile switching center, and interconnection.

1.5.1 Mobile phones

Mobile phones are the portable radio devices that customers use to communicate with the cellular or PCS system. Mobile phones may be mounted in cars or remain as self-contained hand-held units. Mobile phones have many industry names that vary by type of radio.

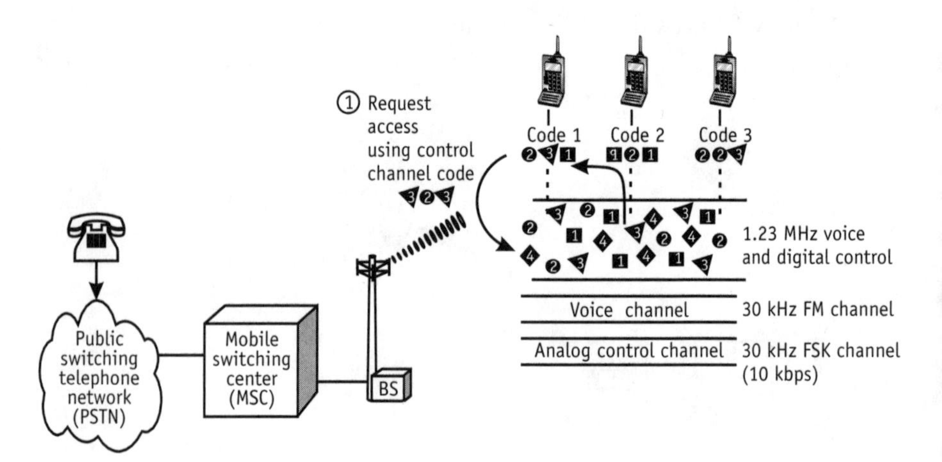

Figure 1.7 CDMA cellular system.

Mobile phones are sometimes called *mobile stations* or *subscriber units.* Hand-held cellular radios may be called *portables;* those mounted in cars may be labeled *mobiles;* and those mounted in bags may be called *bag phones.* This book will refer to any type of mobile cellular radio as a mobile phone. Chapter 4 provides a detailed description of mobile phones.

A mobile phone contains a transceiver (combined radio transmitter and receiver), user interface, and antenna assembly. The transceiver converts audio to RF and RF into audio. The keypad communicates the customer's commands to the transceiver, and its display informs the customer about the phone's operation. The antenna assembly focuses and converts RF energy for transmission and reception. Figure 1.8 presents a block diagram of a typical mobile phone.

To operate correctly in cellular systems, mobile phones must conform to industry-accepted specifications. AMPS mobile phones must conform to Electronic Industries Association 553 (EIA-553)—the "Mobile Station-Land Station Compatibility Specification," 1990—and EIA interim standard 19-B (IS-19-B)—"Recommended Minimum Standards for 800-MHz Cellular Mobile Phones", May 1988. These standards contain parameters for the RF channel structure, signaling, and call processing functions designed to ensure compatibility with the cellular system network. TDMA mobile phones must conform to IS-136 and the minimum performance standard IS-137.

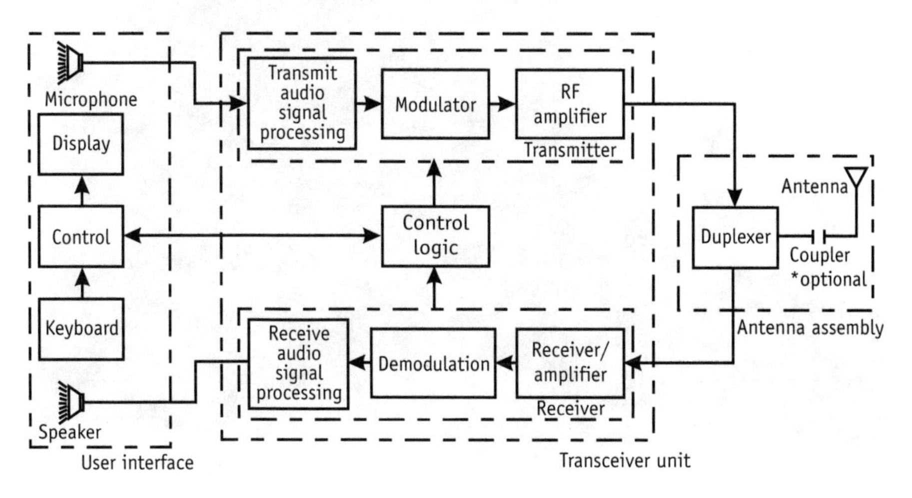

Figure 1.8 Mobile phone block diagram.

Mobile phones are typically classified by power output. Class I mobile phones are capable of 6 dBW (3W), class II phones deliver 2 dBW (1.6W), and class III phones (portables) provide 2 dBW (.6W) of power. Class III units conserve the portables' limited battery storage and reduce emissions by allowing the base station to send commands to adjust the unit's power output.

During the past 10 years, the evolution of mobile telephones has annually reduced the phones' weight by over 80%, their size by more than 90%, and their cost by more than 95%. Figure 1.9 shows the progression of mobile telephones over the last few years.

Mobile phones can be used to connect various other telephone devices. In some locations, it is not practical to connect telephones via wires. In these areas, it is possible to use a cellular telephone and a converter box that allows standard telephones to be connected to a cellular

Figure 1.9 Evolution of cellular telephones. (*Source:* Ericsson.)

telephone. Figure 1.10 shows a converter box that uses a cellular radio to connect it to the telephone system.

One of the new advances in cellular and PCS telephone technology is the use of mobile phones in a local service area. This application may be employed for home cordless service or a wireless office telephone system.

Figure 1.10 Dialtone interface. (*Source:* Telular.)

1.5.2 Base station

Base stations are the radio portion of the wireless system. They consist of multiple radio transceivers, control sections, a radio signal combiner, communications links, a scanning receiver, backup power supplies, and an antenna assembly. The transceivers convert information (voice data or control signaling) to and from radio signals. The control sections coordinate the base station's overall operation. The RF combiner allows the separation of radio channels between multiple transceivers and the antenna assembly. Communications links route audio and control information between the base station and MSC. The scanning receiver measures signal strength on a mobile phone's radio channel to decide when to transfer the calls (hand-off) to other base stations that may have a stronger radio signal. The backup power supply maintains operation when primary power is interrupted. Most sections of the base station are duplicated to maintain functioning if equipment fails. Chapter 5 describes base stations in detail. Figure 1.11 shows a block diagram of a base station.

Base stations are typically located in small buildings adjacent to the radio tower. They usually contain racks of electronics equipment and

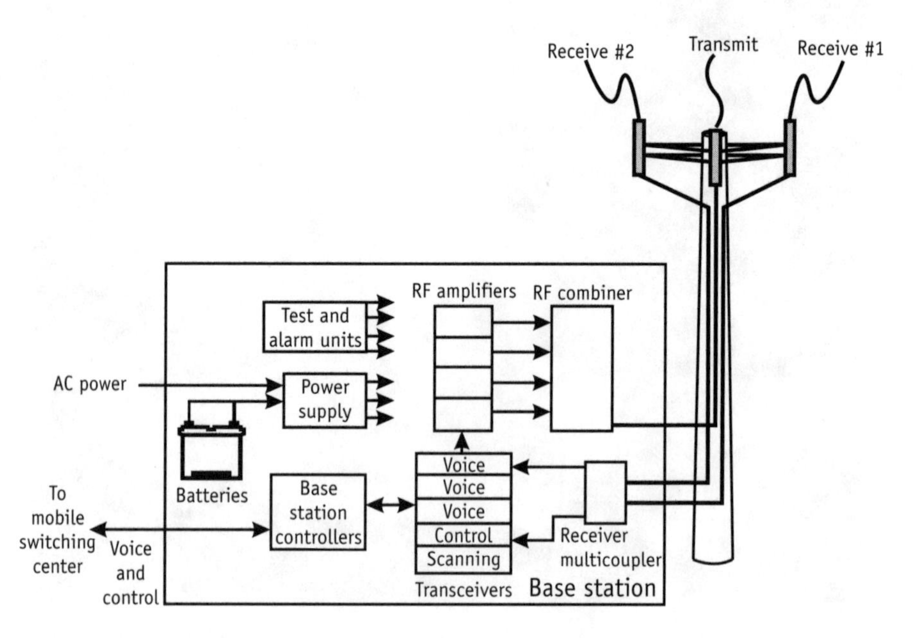

Figure 1.11 Cellular base station block diagram.

require power supply and cooling systems. Figure 1.12 shows typical base station radio equipment.

The trend has been to reduce the size of base stations so they can be mounted on telephone poles or in office buildings. Very small base stations are typically called *microcells.* Microcell base stations are sometimes combined with their antenna assemblies. Figure 1.13 depicts a microcell base station.

1.5.3 Mobile switching center

A cellular or PCS system has a number of MSCs that allow for the connection of mobile phones to other mobile phones or to the wired telephone system. The typical MSC consists of a switching assembly, a controller, communications links, an operator terminal, a subscriber database, and backup energy sources. A switching assembly routes voice connections from the base stations to the PSTN landlines. Communications links connect base stations to the MSC and the MSC to the PSTN. These communications links may be copper wire, microwave, or fiber optic. An operator terminal supervises and maintains the cellular or PCS system. A customer database contains billing records and records of features that customers

Figure 1.12 Base station. (*Source:* Ericsson.)

Figure 1.13 Microcell base station. (*Source:* Hughes Network Systems.)

have specified for their cellular service. Backup energy sources power the MSC during primary power interruptions. Like the base station, the MSC has duplicate circuits and backup energy sources to maintain the system during failures. Figure 1.14 presents a block diagram of an MSC.

Early mobile switching systems were very large and required buildings to house their electronics. Advances in microelectronics have allowed a reduction in the size of switching systems.

A new trend for the switching network is decentralization of the switching system. This allows several miniature switching systems that have autonomous operation to coordinate the access of mobile phones to the cellular system. An example of such a system is a wireless office system that has its own switching system that is not directly connected to the MSC. Chapter 5 provides a detailed description of system networks.

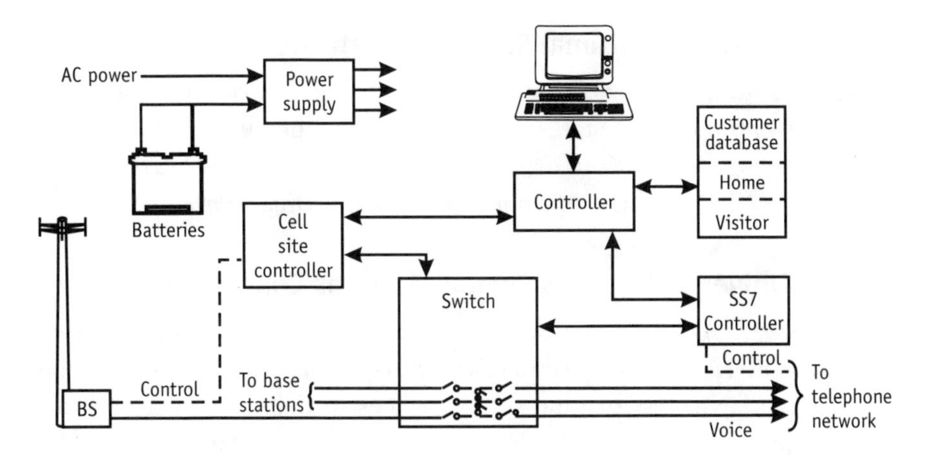

Figure 1.14 MSC block diagram.

1.5.4 Network interconnection

The physical network elements of a cellular or PCS system are interconnected in order to route calls, transfer subscriber information and authenticate users prior to system access. Industry standard communications protocols are used between these devices since different manufacturers supply the various network elements.

1.6 Industry standard groups

Industry standard groups help to create and coordinate common standards that allow equipment and services from different companies to correctly operate with each other. The industry standard groups associated with IS-136 technology include: TIA, ANSI, and UWCC.

1.6.1 Telecommunications Industry Association

The *Telecommunications Industry Association* (TIA) is the part of the EIA that is accredited to the *American National Standards Institute* (ANSI). The TIA is responsible for the creation and updating of standards for telecommunications equipment and systems. This includes the standards associated with analog (AMPS) and digital (IS-136 TDMA) standards. The specific standards associated with IS-136 TDMA are listed in Appendix A.

1.6.2 American National Standards Institute

ANSI is a nonprofit organization that coordinates voluntary standards activities in the United States. ANSI represents the United States in two global telecommunications organizations: the *International Electrotechnical Commission* (IEC) and the *International Standards Organization* (ISO).

1.6.3 Universal Wireless Communications Consortium

The *Universal Wireless Communications Consortium* (UWCC) program was launched in 1995 as a collaborative effort among leading vendors and operators of wireless products and services to deliver enhanced global mobile phone services. These services are independent of radio frequency, market, and customer groups.

The platform for developing and delivering such enhanced personal communications features to customers worldwide consists of the EIA/TIA IS-136 standard version of TDMA radio frequency features, in combination with the enhanced internetworking and mobility of the IS-41 *wireless intelligent network* (WIN) standard capabilities. This provides a complete "engineering tool kit" for provisioning a diversity of public and private wireless service solutions to customers.

To date, UWC conference activity has focused on the market assessment of TDMA/WIN and the coordination of public relations activity. The UWC has created a business organization model to sustain its early success and momentum to market. Formation of the *UWC Consortium* (UWCC) was a commitment to sponsor and manage the UWC conferences.

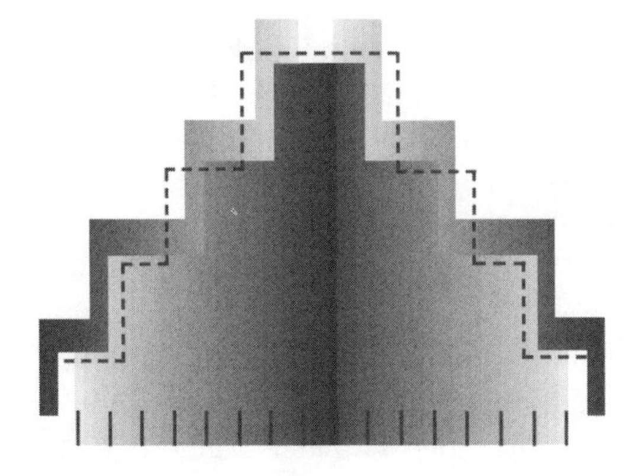

2

Analog Cellular Technology

In 1979, the first U.S. cellular test system began field trials in New Jersey. After four years of testing and refinement, the commercial version of this cellular system was launched in Chicago in October of 1983. This system, AMPS, used analog radio transmission for voice communications developed and patented by AT&T. The original AMPS specification evolved to become IS-3D. Eventually, after several revisions, the IS-3D interim specification became an official EIA specification, EIA-553. This analog cellular system specification is maintained by the TIA working group 45.1.

While cellular service providers (carriers) are permitted to offer any type of cellular technology in their service area, the FCC requires that all 800-MHz cellular carriers offer the standard AMPS cellular service as part of their system. This has resulted in universal availability of AMPS cellular service throughout the United States and over 50 countries internationally.

The U.S. cellular telephone communications network is comprised of cellular carriers operating in 734 different markets. Each market has two cellular license holders: A-system and B-system operators.

To allow for wide area roaming, the new digital cellular standards typically allow for dual-mode mobile phone operation. This allows the mobile phone to access the AMPS or digital system, whichever is preferred. To enable dual-mode operation, the analog cellular specification is incorporated as part of the digital standard (such as IS-136 TDMA), and some modifications to the analog specifications are made (to allow services such as hand-off from an analog channel to a digital channel). These modifications include new security features, control commands, and limited advanced services such as caller identification. Figure 2.1 illustrates the evolution of the analog cellular system specification.

2.1 Functional description

The AMPS cellular system consists of three main parts: mobile phones, base stations, and an MSC. The MSC was formerly known as the *mobile telephone switching office* (MTSO). The mobile phone links the user to base stations via FM radio transmission. Base stations convert signals from mobile phones to signals that are relayed to the MSC. The MSC connects the cellular call to the landline network or other mobile phones.

When an AMPS cellular system is first established, it effectively serves a given number of subscribers, but as the number of subscribers grows, the system must increase its capacity. Adding cells with smaller coverage areas enables a carrier to serve more subscribers in a given area. All radio channels cannot be used in all cells, because using the same frequencies in adjacent cells causes interference and degrades the quality of service.

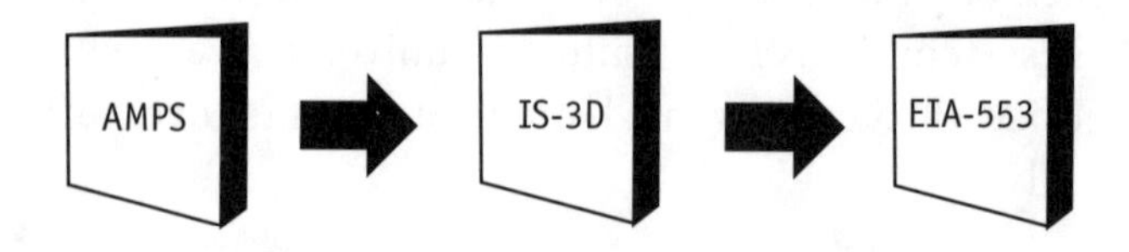

Figure 2.1 Evolution of the AMPS specification.

Mobile phones are comprised of four major elements: a transceiver, a control section, a user interface, and an antenna assembly. These elements are integrated to form a portable mobile phone. The transceiver converts the audio and control information to radio frequencies and converts radio frequencies to audio and control information. A user interface provides the customer with status display information and the ability to command the mobile phone via a keypad or other means. An antenna assembly converts RF between the transceiver and electromagnetic waves propagated in free space.

Base stations contain transceivers, control sections, RF combiners, communications links, scanning receivers, backup power supplies, and antenna assemblies. Several transceivers are provided for voice channels, and one is dedicated as a control channel. Control sections interpret, create, and buffer command and voice information. RF combiners allow a group of transmitters and receivers to share an antenna assembly. Communications links transport voice and data information between the base station and MSC. A scanning receiver measures the strength of the mobile phone signal, which determines when hand-off is possible. Backup power supplies enable the base station to operate when primary power is interrupted. Antenna assemblies convert and direct RF energy between base station equipment and free space electromagnetic waves.

The MSC consists of controllers, a switching assembly, communications links, operator terminals, and backup energy sources and may contain a subscriber database. MSC controllers coordinate all the elements of the cellular system. A cell site controller directs base station equipment; a call controller performs administrative functions; and a communications controller transforms and buffers voice and data information among the MSC, base stations, and PSTN. The switching assembly routes calls between base stations and the PSTN. Communications links transport voice and data via copper wire, microwave, or fiber optic mediums. Operator terminals allow administrative and maintenance information to be observed and entered into the MSC system. A subscriber database contains billing and subscriber preference information. Backup energy sources power the cellular system when primary power is interrupted. Figure 2.2 shows an overview of the AMPS cellular system.

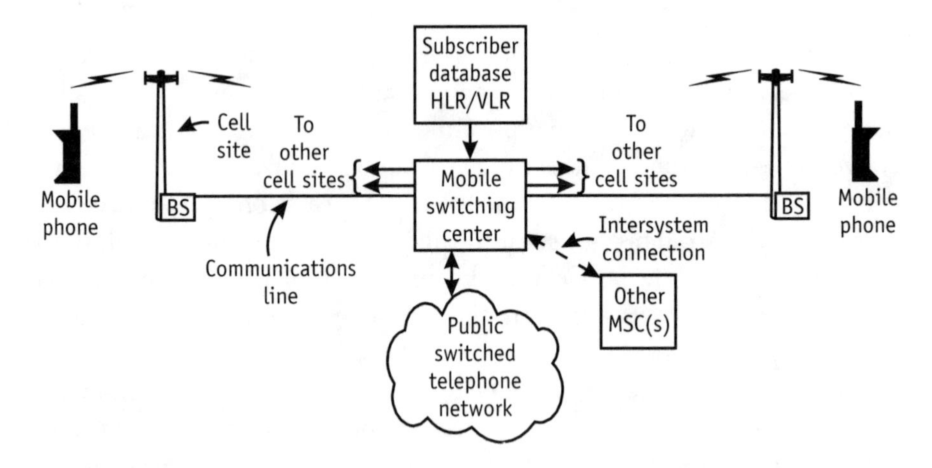

Figure 2.2 AMPS cellular system.

2.2 Analog technology

Analog cellular technology uses various voice and control signal processing, modulation, and RF amplification control technologies. Each of these technologies works together as part of a uniform system that is primarily controlled by the cellular network.

2.2.1 Voice signal processing

Mobile phones convert acoustic voice signals to audio signals and audio signals to acoustic voice signals. The audio signal is called the base band signal. To effectively communicate through a radio channel that is subject to radio signal distortions, the audio signal is processed to minimize the effects of various types of distortions and to exploit the benefits of FM modulation. This processing includes bandpass filtering, companding/expanding, amplitude limiting, and pre-emphasis/de-emphasis.

Bandpass filtering limits the input frequencies to the audio band of 300 to 3,000 Hz. This filter band limits speech signals so that out-of-band speech signals do not influence the gain of the compressor or affect the gain of the amplitude limiter. Figure 2.3 shows how the audio section filters (rejects) the high and low frequencies.

For FM modulation, the amount of frequency shift of the radio signal is proportional to the amplitude of the audio signal. This means that for

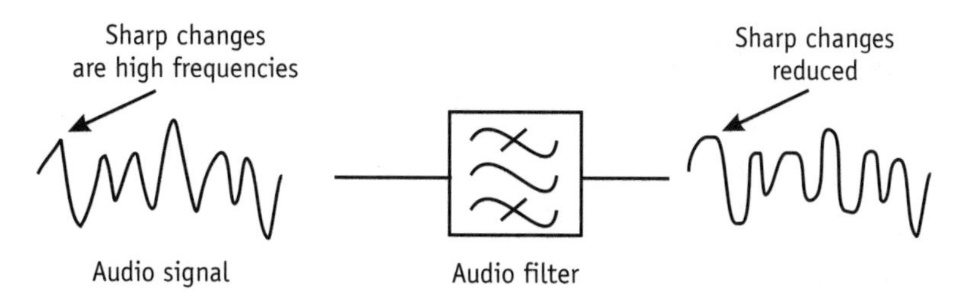

Figure 2.3 Audio filtering.

FM, a quiet speaker will cause a much smaller frequency deviation than a speaker that has a loud voice. To reduce the dynamic range required for FM modulation, a compandor is used to compress the dynamic range of the audio so that speakers with different voice intensities have approximately the same amplitude level that is supplied to the modulator. The compandor varies the gain of the input audio amplifier so that an increased audio signal has a reduced amount of amplification. When the audio signal is received, an expandor varies the gain of the output audio amplifier so that large audio signals receive more gain.

An audio amplitude limiter is required to ensure that an audio signal will not overmodulate the radio signal. Because the offset frequency of FM signals is proportional to the amplitude of the audio signal, a large audio signal would cause the frequency of the radio signal to go outside its authorized frequency bandwidth. This would likely interfere with users of the adjacent radio channel. The amplitude limiter reduces the gain of the input audio amplifier dramatically if the input audio signal exceeds its limits. Figure 2.4 shows the audio limiting process.

Speech audio signals contain most of their energy in the low frequencies. Thus, the signal-to-noise ratio of the audio signal at higher frequencies is lower. The signal-to-noise ratio of high-frequency components is further degraded as it is typical for the noise output of the demodulator to increase exponentially at higher frequencies. Therefore, the noise output of the demodulator is highest where the audio signal frequency components are the smallest.

To overcome the limitations of the high-frequency portions of the audio signal, a pre-emphasis section is used to optimize the high-frequency components of the audio signal for FM transmission. The pre-emphasis section varies the gain of the input audio amplifier so that an

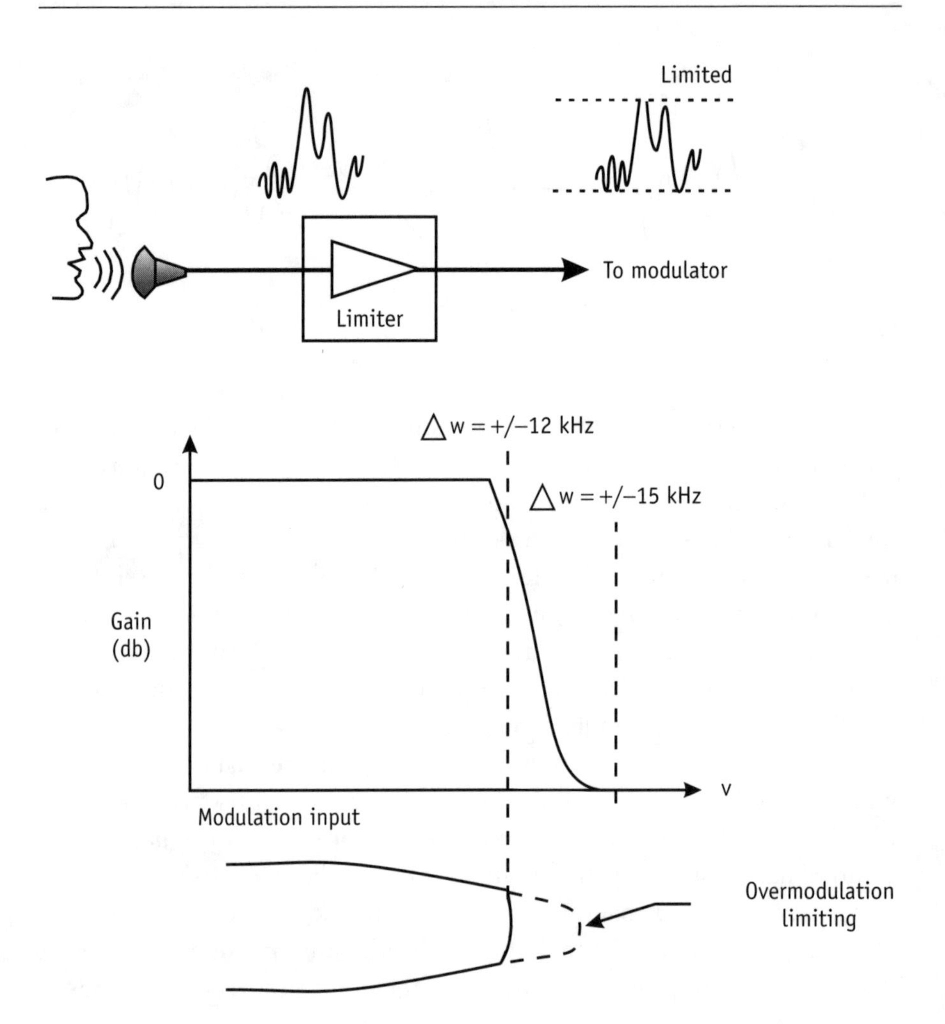

Figure 2.4 Amplitude limiting.

increased audio frequency has an increased amount of amplification. When the audio signal is received, a de-emphasis section varies the gain of the output audio amplifier so that high-frequency audio signals receive less gain.

2.2.2 Channel multiplexing

A channel multiplexer allows the insertion of control tones into the audio path. This enables the system to verify the communications link with the phone and to avoid cochannel issues when two phones are using the same frequency in different cells. One of several 6-kHz *supervisory audio tones* (SATs) is continuously combined with the audio signal during transmission. A 10- kHz *signaling tone* (ST) is sometimes combined with the audio signal depending on the signaling condition requirements. Because these control tones are higher than the 3-kHz audio limit, they are combined after the audio bandpass filter. When these tones are received, they are blocked from the received audio signal by another audio bandpass filter. Figure 2.5 depicts the channel multiplexing process.

2.2.3 Modulation

The analog cellular system uses FM modulation to transfer voice information and FSK modulation to transfer digital information. RF modulation converts information into a radio signal. The modulation input signal containing information is the baseband signal. The RF carrier that transports the information is the broadband signal.

FM is a process of shifting the radio frequency in proportion to the amplitude (voltage) of the input signal. FM is not as susceptible to noise and varying channel conditions as AM signals are. This is because the information contained in an FM signal is proportional to the frequency change of the carrier signal. For an FM signal, without modulation input, the output of the modulator is a constant carrier frequency (ω_c). As the

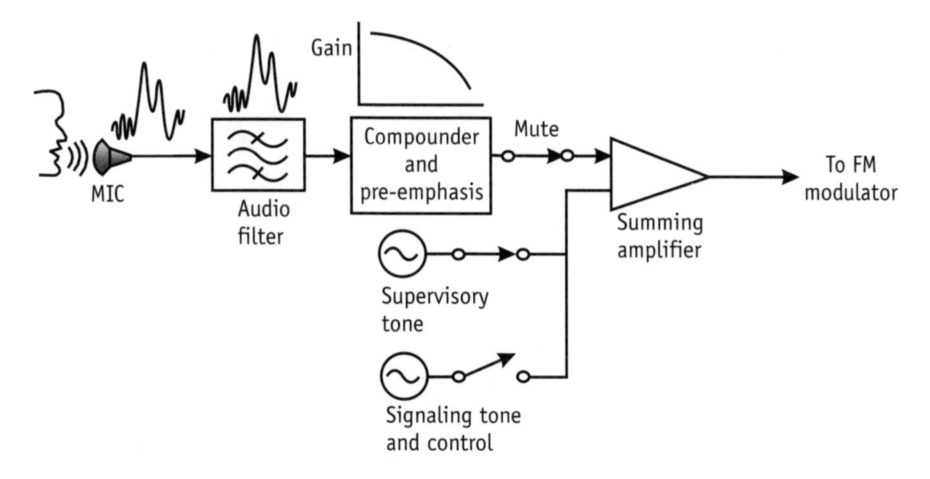

Figure 2.5 Channel multiplexing.

audio (information) signal is applied to the modulator, the carrier frequency is offset in proportion to the amplitude of the information signal. The maximum amount of offset the carrier can be shifted above or below the reference is called the peak deviation. Generally, positive and negative peak deviations are equal. Figure 2.6 shows how FM modulation is used for the AMPS mobile phone.

Most AMPS phones use phase modulation that converts a voltage level input to a proportional frequency shift of the carrier signal. To

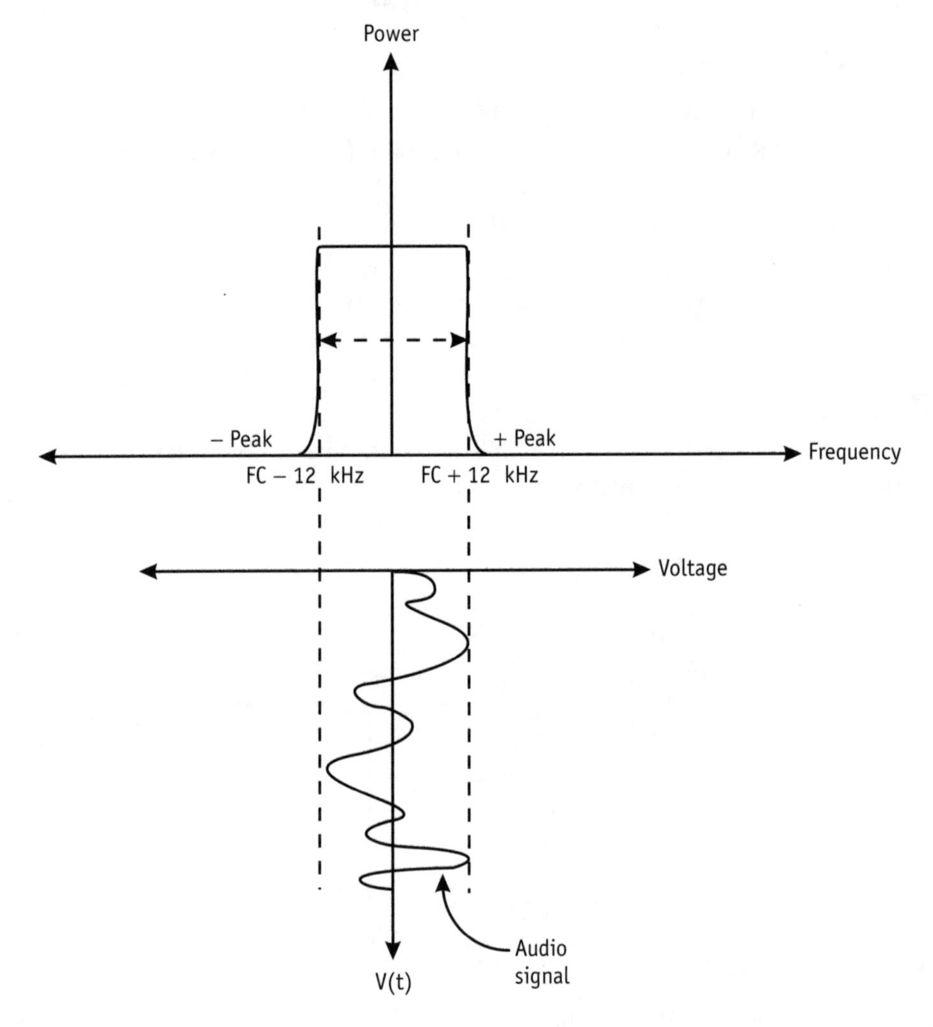

Figure 2.6 FM modulation.

optimize modulation for the mobile environment operating at 800-MHz radio, special audio processing enhances high frequencies and minimizes transients caused by 800-MHz radio signal fades.

Analog cellular systems transfer some messages via digital signaling. To send digital messages, the digital information can be transmitted using FM modulation. When a digital signal is applied to an FM modulator, the output frequency changes depending on the level of the input signal.

2.2.4 RF amplification

To help reduce interference from a mobile phone's transmitted radio signal to nearby mobile phones and base stations, the mobile phone's output power level is continually adjusted so that it only transmits the necessary power level to reach its serving base station. To control this process, the serving base station detects the received power level and then sends commands to the mobile phone adjusting its power level up or down as necessary. Figure 2.7 shows that the RF power amplifier can vary its output power based on commands received from the base station.

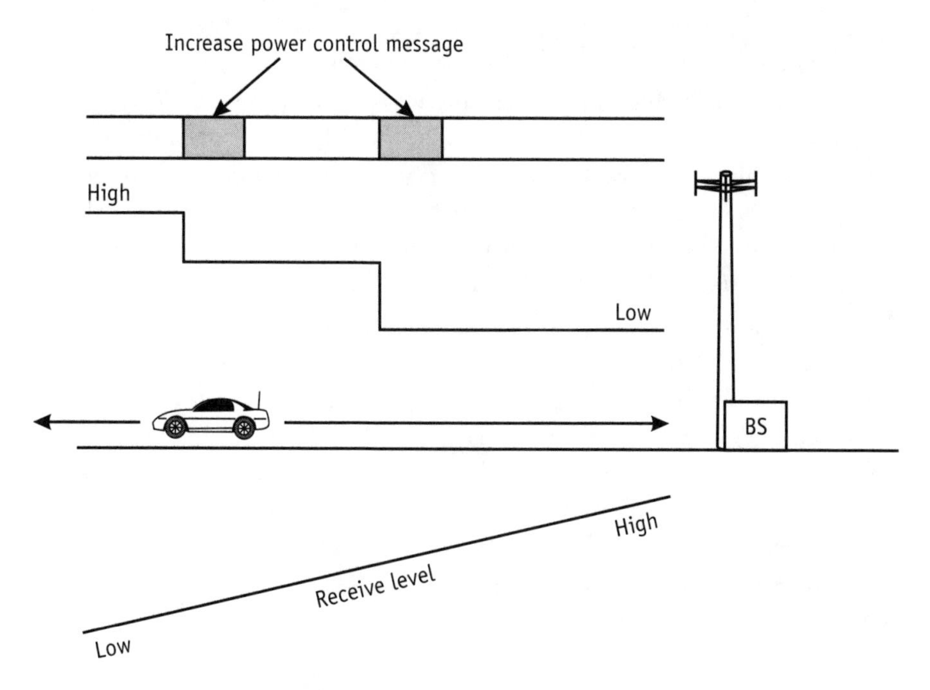

Figure 2.7 RF power level control.

Because the AMPS cellular system uses FM modulation, distortions created by the RF amplifier that change the shape of the radio signal (such as clipping the top of the signal) do not typically reduce the audio quality of the signal. This allows AMPS cellular telephones to use efficient class C RF amplifiers whose efficiency comes at the expense of added distortions to the shape of the radio signal. The RF energy conversion efficiency for class C RF amplifiers in AMPS phones is typically above 55%.

2.3 System parameters

The cellular system was developed with the technology constraints of the 1970s. Frequency bands suitable to cost-effective equipment design were limited, the selected modulation type had to conform to a hostile mobile radio environment, and the control structure had to support large numbers of available voice channels.

2.3.1 Channel frequency allocation

To provide cellular service, companies in the United States must obtain a cellular license from the FCC. The FCC's licensing system is designed to support many users with limited frequency spectrum. In the 1970s, available radio spectrum below 1 GHz was limited, and equipment design limitations and poor radio propagation characteristics at frequencies above 1 GHz led the FCC to allocate 825–890 MHz for cellular radio. This frequency band came from the original frequency band allocated for *ultra high frequency* (UHF) television. The initial allocation was for 40 MHz, and in 1986, an additional 10 MHz of spectrum was added to facilitate expansion.[1]

Cellular service areas are shared between two cellular companies, called A and B carriers. Today, the United States is divided into 73 cellular service areas, each with an A and a B carrier. The A carrier does not have a controlling interest in the local telephone company, while the B carrier (often a Bell operating company) can have a controlling interest in a local telephone company. The A and B carriers are each licensed to use 25 MHz of radio spectrum. Each carrier divides its

1. William Lee, *Mobile Cellular Telecommunications Systems,* McGraw Hill, 1989, p. 265.

25 MHz into no more than 416 radio channels. Twenty-one of these channels are used for control and paging, and 395 are used for voice. Figure 2.8 illustrates cellular channel allocations.

To calculate the frequency of the radio channel, the following formulas, in which N represents the channel number, are used:

- Reverse channel:

  1 to 799: $0.03(N) + 825$ MHz;

  990 to 1023: $0.03(N - 1023) + 825$ MHz.

- Forward channel:

  1 to 799: $0.03(N) + 870$ MHz;

  990 to 1023: $0.03(N - 1023) + 870$ MHz.

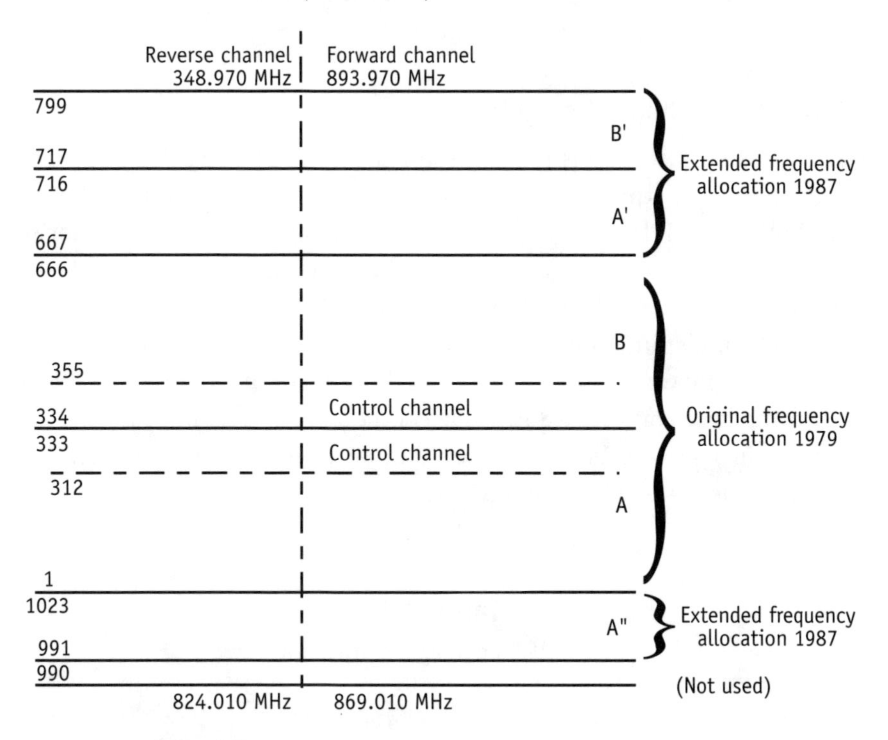

Figure 2.8 AMPS channel frequency allocation.

2.3.2 Radio frequency power levels

Mobile phones are classified by maximum RF power output. AMPS cellular phones have three classes of maximum output power. Class 1 maximum power output is 6 dBW (4W), Class 2 is 2 dBW (1.6W), and Class 3 is −2 dBW (.6W). However, actual mobile phones' RF power outputs vary, because base station commands adjust output in increments of 4 dB down to the minimum for all AMPS mobile phones of −22 dBW (6 mW). Table 2.1 shows the AMPS mobile phone RF power classifications.

2.3.3 Modulation limits

The modulation limit for an AMPS phone is +/− 15 kHz to contain the radio signal in the 30 kHz radio pass band. The peak deviation of the carrier signal is a result of the combination of the audio (voice) and control (SAT and ST). Since SAT is continuously transmitted during conversation, the maximum deviation due to audio is +/− 12 kHz.

2.3.4 Radio channel structure

To support simultaneous transmission and reception (called *frequency duplex operation*), the base stations transmit on one set of radio channels (869–894 MHz), called *forward channels,* and receive on another set (824–849 MHz), called *reverse channels.* Forward and reverse channels in each cell are separated by 45 MHz. Figure 2.9 illustrates forward and reverse channels. In Figure 2.9(a), a base station transmits to the mobile phone on the forward channel at 875 MHz. The mobile phone then transmits to the base station at 830 MHz on the reverse channel. If, alternatively, as in Figure 2.9(b), the base station transmits at 890 MHz, the mobile phone transmits at 845 MHz.

Table 2.1

AMPS RF Power Classification

RF Power	Class 1	Class 2	Class 3
Maximum power	6 dBW (4W)	2 dBW (1.6W)	−2 dBW (.6W)
Minimum power	−22 dBW (6 mW)	−22 dBW (6 mW)	−22 dBW (6 mW)

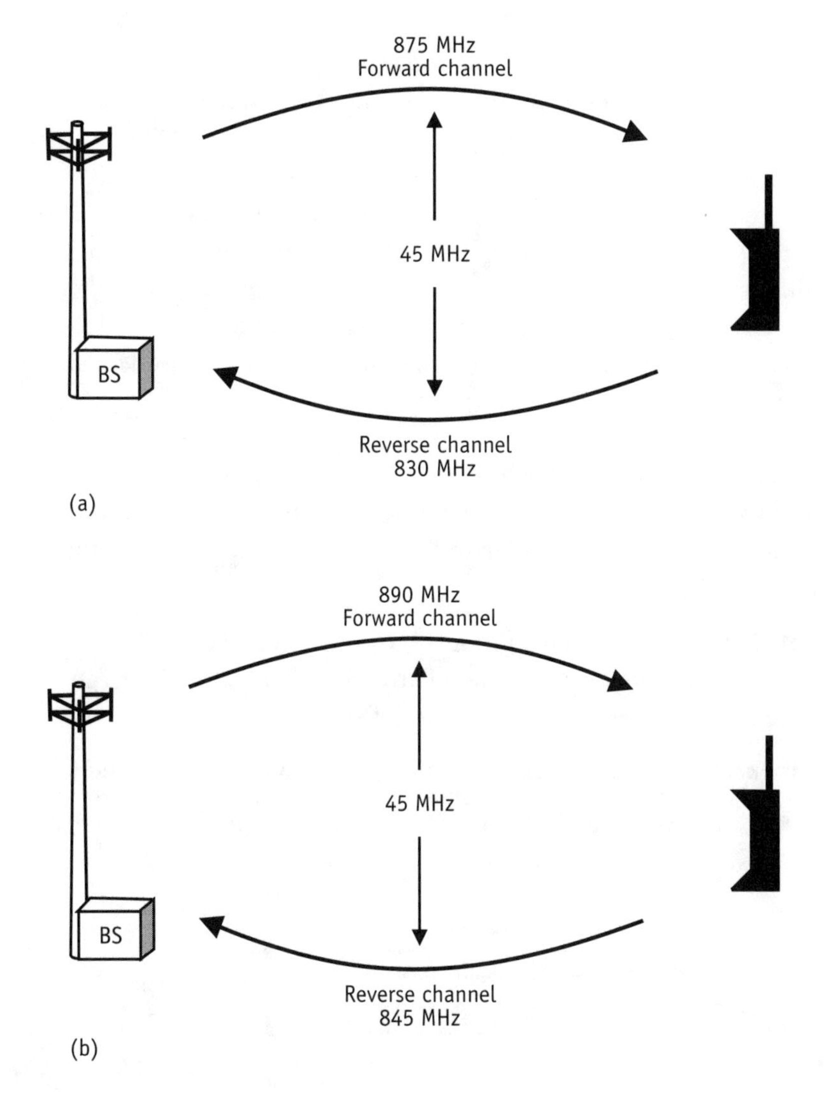

Figure 2.9 Duplex radio channel spacing.

In early mobile radio systems, a mobile phone scanned the few available channels until it found an unused one, but today, no mobile phone could scan the 832 channels in a reasonable amount of time.[2] Therefore,

2. *The Bell System Technical Journal,* Vol. 58, No. 1, Murray Hill, New Jersey: American Telephone and Telegraph Company, January 1979, p. 50.

control channels were established to quickly direct a mobile phone to an available channel. Today, of the 416 channels available per system, 21 are control channels and cannot be used as voice channels. The remaining 395 channels are either voice or control channels.

A mobile phone communicates with the cellular system by sending signaling messages. Signaling message formats vary between control channels and voice channels. Control channel signaling is all digital and voice channel signaling is a mixture of digital messages and audio tones. Figure 2.10 shows the different types of radio channels used on an AMPS system.

2.3.4.1 Analog control channels

The dedicated analog control channels carry the following four types of messages to allow the cellular radio to listen for pages and compete for access:

- *Overhead messages* continuously communicate the *system identification* (SID) number, power levels for initial transmissions, and other important system registration information.
- *Pages* tell a particular cellular radio that a call is to be received.
- *Access information* is exchanged between the mobile phone and the system to request service.

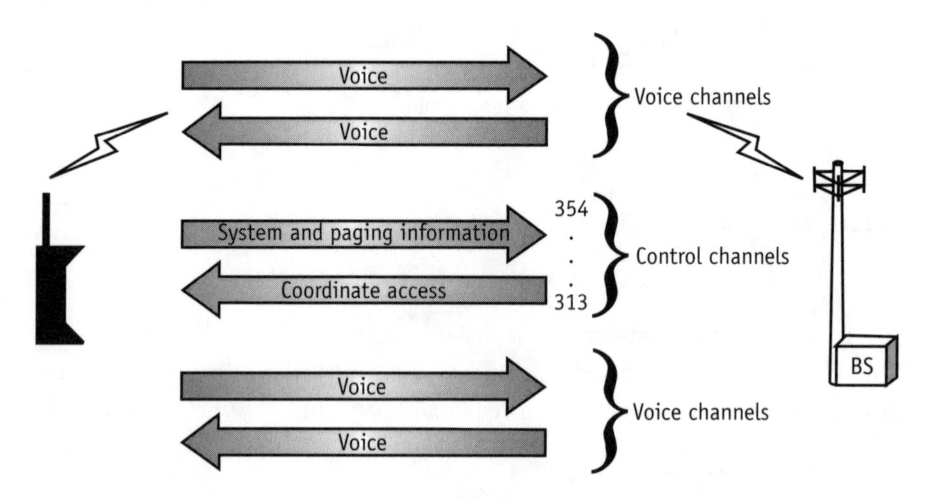

Figure 2.10 AMPS radio channel structure.

- *Channel assignment commands* establish the radio channels for voice communications.

2.3.4.2 Voice channels

After a mobile phone is assigned a voice channel, FM signals transmit the voice, and FSK signals transmit control messages. Control messages that are sent on the voice channel include the following:

- *Hand-off messages,* which instruct the mobile phone to tune to a new channel;
- *Alert,* which tells the mobile phone to ring when a call is to be received;
- *Maintenance commands,* which monitor the status of the mobile phone;
- *Flash,* which requests a special service (such as three-way calling) from the system.

2.4 System security

A cellular system must be capable of uniquely identifying a mobile phone and determining that it is authorized to use the services of the cellular network. To uniquely identify mobile phones, all AMPS phones are manufactured with a unique *electronic serial number* (ESN). The ESN is designed so that it is not readily alterable by the customer. In addition to the ESN, a mobile phone is assigned a *mobile identification number* (MIN) by the customer's cellular service provider. The MIN is typically the customer's telephone number. The MIN and other system specific information is stored in the *number assignment module* (NAM) memory area inside the mobile phone. Typically, the retailer that sells the telephone to the customer programs the NAM.

The combination of the ESN and the MIN allows the cellular system to determine the identity of the mobile phone. Unfortunately, criminals have used duplicate (cloned) phones fraudulently with the ESN and MIN of valid paying customers to get free calls. This has necessitated the addition of better system security. The cellular system has evolved to include authentication, which is used to confirm the identity of a mobile phone by using secret stored information to validate transferred information. Figure 2.11 shows the authentication process.

2.5 Signaling

Signaling is the process of transferring information messages. The AMPS control channel sends information messages by FSK at a rate of 10 Kbps. To allow self-synchronization, the information is Manchester encoded, forcing a frequency shift (bit transition) for each bit input. Orders are sent as messages of one or more words.

2.5.1 Forward control channel

The forward control channel transfers control messages from the base station to the mobile phone. Messages are continually sent on the forward control channel. A mobile phone must first synchronize to the message structure on the control channel. To initially learn when the beginning of a new message is coming, a dotting sequence of alternating 1's and 0's is sent on the control channel to indicate that a message is about to begin. The alternating bits produce a strong, easily detected 5-kHz frequency

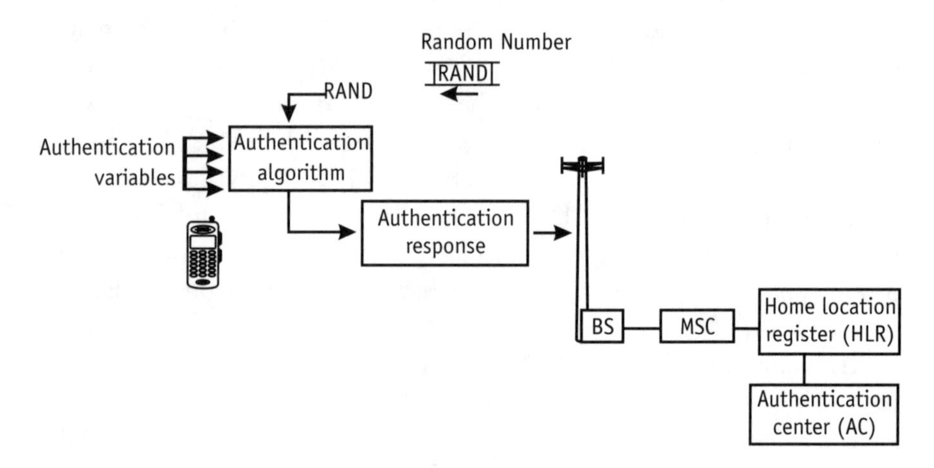

Figure 2.11 Authentication.

component.[3] A synchronization word follows the dotting sequence to define the exact starting point of the coming message. Message words follow the synchronization word. Since radio channels can fade and introduce errors, the message words are repeated five times to ensure reliability. Of the five repeats, a majority vote of three out of five words can be used to eliminate corrupted messages. On the forward control channel, ten words follow the dotting/sync word sequence. Words are alternated A, B, A, B, etc. A words are designated for mobile units with even phone numbers. B words are designated for mobile phones with odd phone numbers. A forward channel word is 40 bits. Each word includes BCH error correction/detection, containing 28 bits with parity of 12 bits. Figure 2.12 shows the signaling structure of the forward control channel.

To help coordinate multiple mobile phones accessing the system, busy/idle indicator bits are interlaced with the other bits. Before a mobile phone attempts access, it checks the busy/idle bits to see if the control channel is serving another mobile phone. This system, called *carrier sense multiple access* (CSMA), helps avoid collisions during access attempts. Figure 2.13 illustrates coordination of access using busy/idle bits.

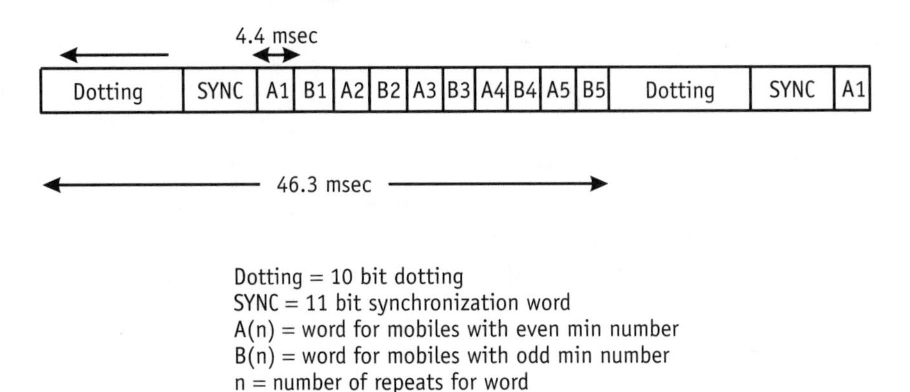

Figure 2.12 Forward control channel signaling.

3. *The Bell System Technical Journal,* Vol. 58, No. 1, Murray Hill, New Jersey: American Telephone and Telegraph Company, January 1979, p. 50.

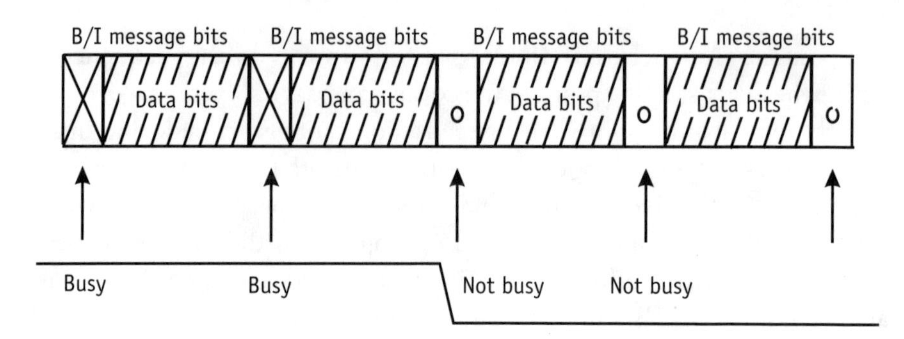

Figure 2.13 Busy idle bits.

2.5.2 Reverse control channel

On the reverse channel, five words follow the dotting sequence. A reverse channel word is 48 bits. Each reverse channel word includes BCH error correction/detection, containing 36 bits with parity of 12 bits. Messages on the reverse channel are sent in random order and coordinated using the busy/idle bits from the forward control channel. Figure 2.14 shows reverse control channel signaling.

2.5.3 Voice channel signaling

The analog voice channel passes user information (usually voice information) between the mobile phone and the base station. Signaling information must also be sent to provide physical layer control. Signaling on the voice channel can be divided into in-band and out-of-band signaling. In-band signaling occurs when audio signals between 300–3,000 Hz either replace or occur simultaneously with voice information. Out-of-band

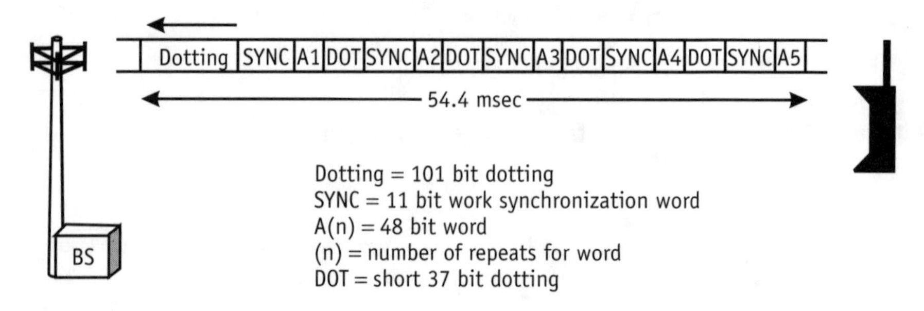

Figure 2.14 Reverse control channel signaling.

signals are above or below the 300–3,000 Hz range and may be transferred without altering voice information.

Signals sent on the voice channel include SAT, ST, *dual-tone multifrequency* (DTMF), and blank and burst FSK digital messages.

2.5.3.1 Supervisory audio tone

The SAT tone is used to verify the reliable transmission path between the mobile phone and base station. The SAT tone is transmitted along with the voice to indicate a closed loop. The tone functions much like the current/voltage used in landline telephone systems to indicate that a phone is off the hook.[4] The SAT tone may be one of the three frequencies: 5,970, 6,000, or 6,030 Hz. A loss of SAT implies that channel conditions are impaired. If the SAT tone is interrupted for longer than about five seconds, the call is terminated. Figure 2.15 shows how a SAT tone is returned (transponded) to a base station to confirm a radio connection is reliable.

SAT can also mute the effects of cochannel interference. Interfering signals have a different SAT frequency than the one designated by the system for the call in progress. The incorrect SAT code alerts the mobile phone to mute the audio from the interfering signal.

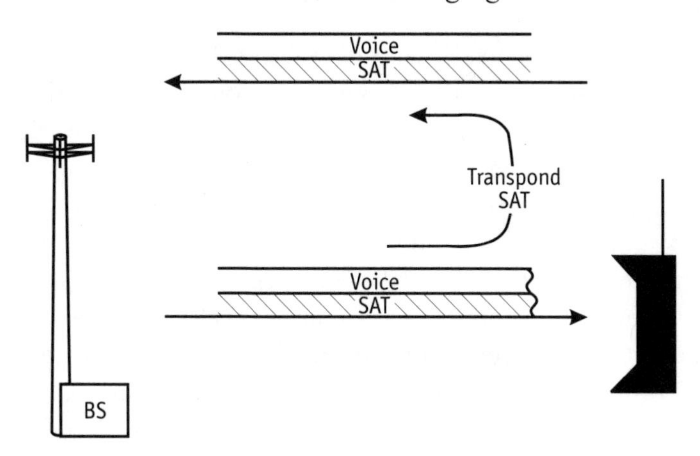

Figure 2.15 Transponding SAT.

4. *The Bell System Technical Journal,* Vol. 58, No. 1, Murray Hill, New Jersey: American Telephone and Telegraph Company, January 1979, p. 47.

2.5.3.2 Signaling tone

The ST is a 10-kHz tone burst used to indicate a status change. It confirms messages sent from the base station and is similar to a landline phone going on or off hook.[5]

2.5.3.3 Dual-tone multifrequency

Touch-tone (registered trademark of AT&T) signals (DTMF) may be sent over the voice channel. DTMF signals are used to retrieve answering machine messages, to direct automated PBX systems to extensions, and to perform a variety of other control functions. Telecommunications industry standards specify the frequency, amplitude, and minimum tone duration for recognition of DTMF tones.

2.5.3.4 Blank and burst messages

When signaling data is about to be sent on the voice channel, audio FM signals are inhibited and replaced with digital messages. This voice interruption is normally too short (34–54 msec) to be heard. The bit rate for messages is 10 Kbps, and messages are transmitted by FSK. Like control channel messages, these messages are repeated and a majority vote is taken to see which messages will be used.

To inform the receiver that a digital signaling message is coming, a 101-bit dotting sequence produces a 5-kHz tone preceding the message. A synchronization word follows the dotting sequence to identify the exact start of the message. Figure 2.16 illustrates how a voice channel message transmission mutes the audio.

Blank and burst signaling differs on the forward and reverse voice channels. On the forward voice channel, messages are repeated 11 times to ensure that control information is reliable, even in poor radio conditions. On the reverse voice channel, words are repeated only five times. Words contain 40 bits on the forward voice channel and 48 bits on the reverse voice channel. Both types of words have 12 bits of BCH error detect/correct parity. Figures 2.17 and 2.18 display how signaling on the voice channel varies between the forward and reverse channels.

---

5. *The Bell System Technical Journal,* Vol. 58, No. 1, Murray Hill, New Jersey: American Telephone and Telegraph Company, January 1979, p. 47.

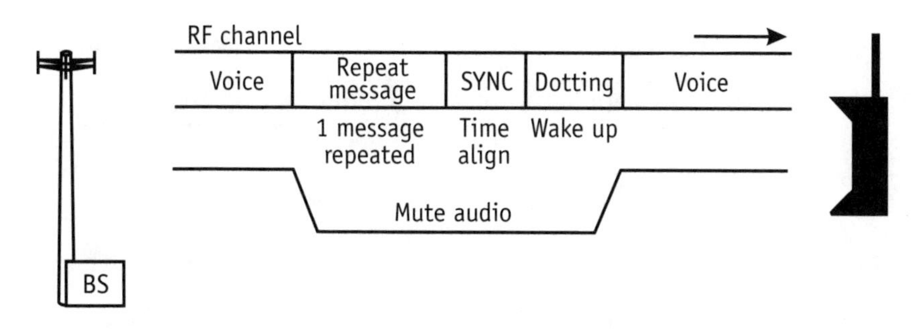

Figure 2.16 Voice channel muting.

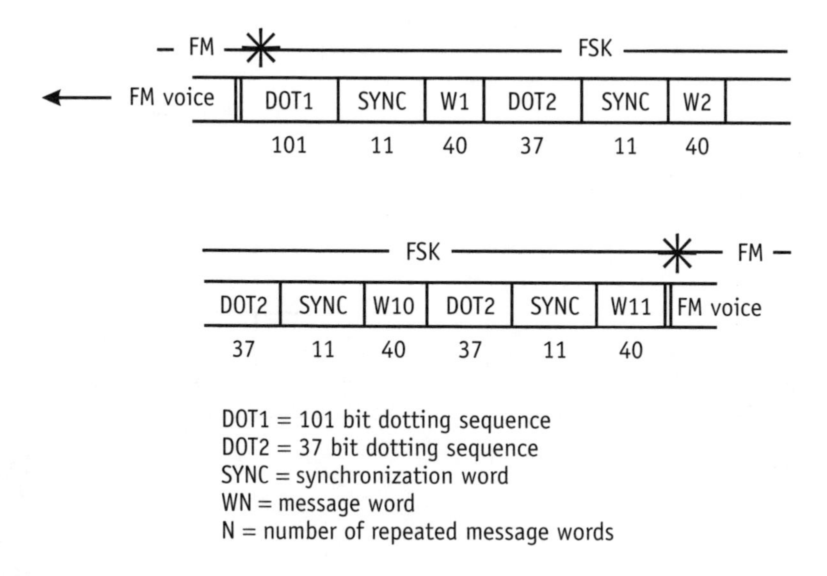

Figure 2.17 Forward voice channel.

2.6 System operation

When a mobile phone is first turned on, it must locate a control channel. It does this by scanning the predetermined set of analog control channels and tuning to the strongest one. The mobile phone then initializes parts of its memory by listening to the messages on the control channel. Figure 2.19 shows that during this initialization mode, the mobile phone retrieves various SID and setup information.

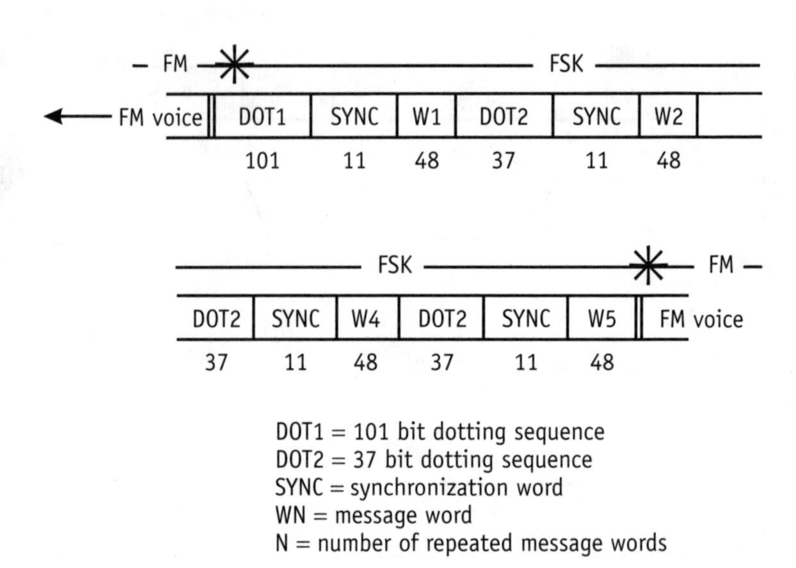

Figure 2.18 Reverse voice channel.

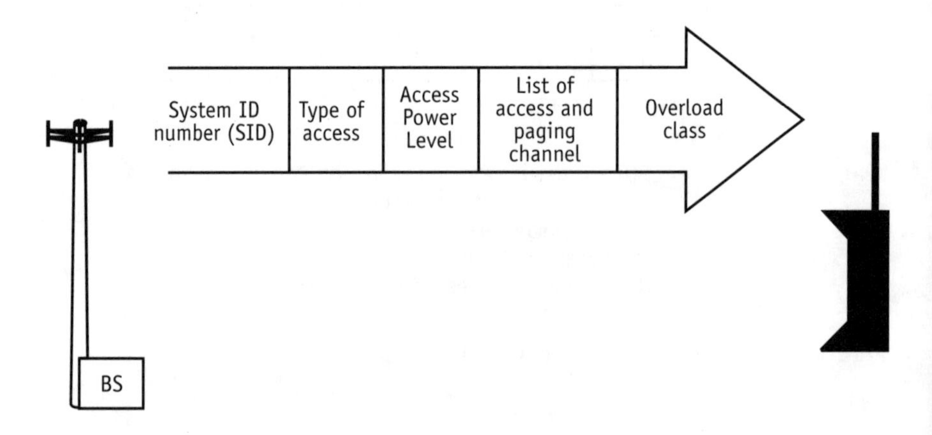

Figure 2.19 Cellular system broadcast information.

2.6.1 Access

After initialization, the mobile phone enters idle mode and waits either to be paged for an incoming call or for the user to place a call. When a call is to be placed, the mobile phone enters the system access mode and originates a call on the strongest analog control channel. If the attempt is successful, the system sends an *initial voice channel designation* (IVCD) message to indicate an open voice channel. The mobile phone tunes to the designated voice channel and enters conversation mode.

2.6.2 Paging

Mobile phones are notified of incoming calls through the paging process. A page is a control channel message containing the MIN (phone number) to indicate that an incoming call is to be received. Pages are sent over a wide area of the system since the location of the phone is not known on a cell-by-cell basis. The phone responds to the page, and the system sends an IVCD message to indicate an open voice channel. In the same way as system access, the phone retunes to the designated voice channel and enters the conversation mode.

2.6.3 Hand-off

As a mobile phone moves away from the cell that is serving it, the cellular system must eventually transfer service to a more suitable cell. To determine when hand-off is necessary, the serving base station continuously monitors the mobile phone's signal strength. When signal strength falls below a minimum, the serving base station requests adjacent base stations to measure that specific mobile phone's signal strength. The adjacent base stations tune to the mobile phone's current operating radio channel and measure the mobile phone's signal strength. When a nearer adjacent base station measures sufficient signal strength, the serving base station commands the cellular radio to switch to the nearer base station. After the cellular radio starts communicating with the new base station, the communications link carrying the landline voice path is switched to the new serving base station to complete the hand-off.

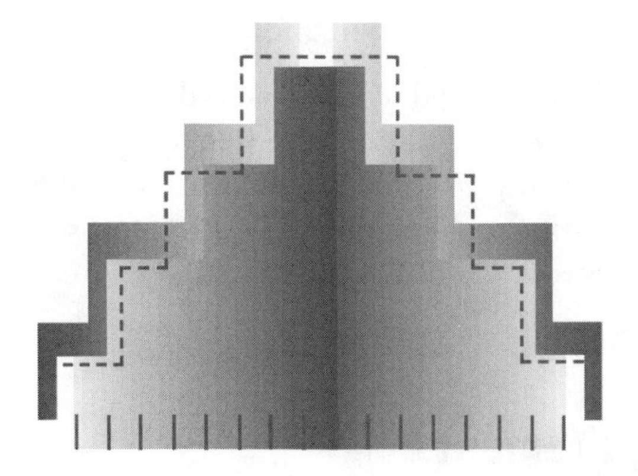

3

IS-136 TDMA Technology

The IS-136 standard is the basis of the TDMA cellular and personal communications services (PCS) air interface technology. IS-136 systems allow telecommunications companies to offer advanced features and services to subscribers while maintaining backward compatibility for their existing customer bases. It also provides a cost-effective migration path from AMPS and earlier TDMA systems through the use of dual-mode operation—where analog or digital capabilities can be used according to system resources and phone capability.

The DCCH forms the core of the IS-136 specification. It is the primary enhancement to IS-54B (ANSI standard TIA/EIA 627) technology and represents the next generation of TDMA-based digital operation. The analog portion of the AMPS EIA-553 specification was incorporated into the new digital specification to provide a smooth migration path and to continue the analog and digital dual mode philosophy.

The IS-136 DCCH makes TDMA a powerful PCS technology since it provides a platform for advanced features and services and is designed to work seamlessly on either the 800-MHz or 1,900-MHz frequency.

The IS-136 specification is actually two documents containing the specification information necessary for development of IS-136-based products. Figure 3.1 shows the evolution of the digital cellular system specification.

3.1 Technology requirements

3.1.1 New features

IS-136 introduces the DCCH, which provides new system functionality and supports enhanced features including the following.

- A battery life power saving process called sleep mode;
- Support for multiple vocoders to take advantage of new voice improvement technology;
- The ability to seamlessly acquire the same services in either the cellular (800-MHz) or so-called PCS (1,900 MHz) frequency band;
- A teleservice feature for transferring application data to and from cellular phones—including the *cellular messaging teleservice* (CMT),

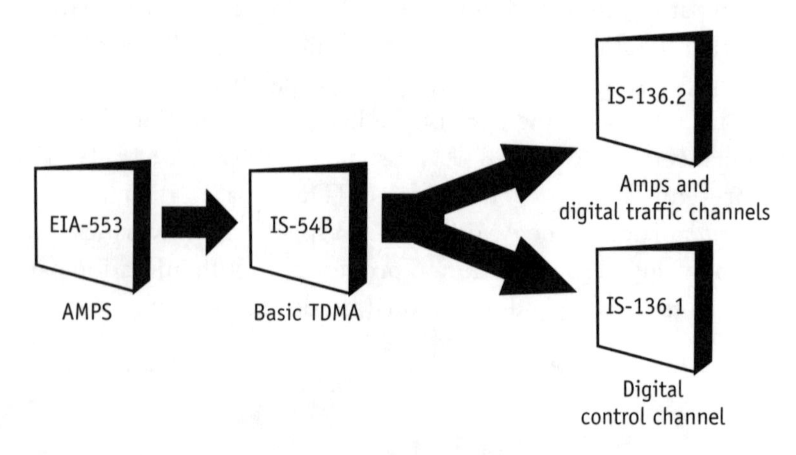

Figure 3.1 Evolution of the IS-136 specification.

which delivers short alphanumeric messages to the phone, and the *over-the-air activation* (OAA) *teleservice* (OATS), which allows for delivery of provisioning information to the phone;

- A hierarchical macrocell-microcell environment providing support for microcellular operation;
- Private and residential system identities providing the tools for *wireless office service* (WOS) operation;
- The ability to quickly roll out advanced services to meet future consumer needs.

These features are discussed fully in Chapter 10.

The *Cellular Telephone Industry Association* (CTIA) commissioned the creation of a *user performance requirements document* (UPR) in 1989. The basic requirements of the UPR, which are supported in the IS-136 environment, include increased capacity, new features, equipment availability, voice quality, and backward compatibility.

3.1.2 Increased capacity

It appeared that cellular systems were running into a capacity limit in the early 1990s. In some urban areas, cell sites were approaching a 1/2-mile radius, which was considered the minimum practical cell size for an AMPS system. This meant that cellular carriers were concerned that there was a fundamental limit on the maximum number of customers that they could serve in a given geographical area. Serving more customers simultaneously with the same amount of radio spectrum was considered essential to continued growth and survival.

There were two capacity issues: voice channel capacity and control channel capacity. Voice channel capacity was increased by dividing the radio channel into time slots, and control channel capacity was increased by adding a DCCH.

3.1.3 New features

To better serve their customers, cellular carriers specified new features such as short messaging, extended battery life, and data services. A priority list was created for key features (such as calling number identification) and low priority features (such as video service).

3.1.4 Equipment availability

The cellular carriers wanted to have digital mobile phones available quickly, to increase capacity, and to allow for advanced features. Since digital-only service was not expected to be available throughout the entire United States for several years, digital equipment was expected to be dual-mode, or capable of operating on both the AMPS and digital system, where appropriate.

3.1.5 Voice quality

Cellular carriers aim to provide voice quality comparable to that supplied by wired telephones. Due to the robust nature of digital communications, digital voice service has, in many cases, voice quality superior to that of AMPS analog technology.

3.1.6 Compatibility with AMPS and IS-54B systems

A primary consideration in early plans for the DCCH and the introduction of its features was the ability to implement the technology quickly and to minimize disruption in the existing cellular system. To achieve this, many of the first generation TDMA (IS-54B) physical radio characteristics are retained in IS-136. For example, in addition to retaining the TDMA slot and frame structure, the IS-54B in-call messaging remains the same in the IS-136 environment. Table 3.1 lists the compatibility and transmission requirements addressed during development of the IS-136 standard.

3.2 IS-136 radio technology

The radio technology used in the IS-136 system provides a channel for advanced services and improved system efficiency through the use of voice digitization, speech compression (coding), channel coding, efficient radio modulation, enhanced RF power control, and a flexible approach to spectrum usage.

Table 3.1
Key System Development Requirements

Requirement	Benefit
Allow a new control channel (DCCH) to coexist with the current TDMA cellular channels	Multiple number of control channels per base station can exist
Maintain the same type of modulation as IS-54B	No significant system hardware changes would be required
Maintain the IS-54B basic call processing process	Simplified equipment design to perform on IS-136 system
Make only minimal changes to the DTC	IS-54B phones will not be confused with advanced features when they operate in the IS-136 system
Keep the same set of channel coding (error protection) types for the DCCH and the DTC	Simplified signal processing reduces complexity and power consumption
Capability to mix a DTC and a DCCH on any frequency	Allows a flexible use of the radio spectrum

3.2.1 Voice digitization

The first step in a digital cellular system is the conversion of an acoustic voice signal into a digital signal. As the customer speaks into the microphone, an analog audio signal is created. The audio signal is very complex and contains very high and low frequencies that are not necessary for communication. A filter is, therefore, used to remove any signals below 100 Hz or above 3,000 Hz before further processing. The filtered audio signal is then converted to a digital value at a sampling rate of 8,000 times per second. For each sample, an eight-bit digital value is created. The resulting 64,000-bps digital signal represents the voice information.

3.2.2 Speech data compression

After the voice digitization process, the digitized audio signal is typically 64 Kbps. To efficiently send a digitized voice signal, the IS-136 uses speech data compression. This is performed by a speech coder. The speech coder characterizes the digitized audio signal and attempts to ignore patterns that are not characteristic of the human voice. The result is a digital signal that represents the voice content. When this compressed speech information is received, a speech decoder is used to recreate the original signal.

The IS-136 speech coder analyzes the 64 Kbps speech information and characterizes it by pitch, volume, and other parameters. Figure 3.2 illustrates the speech compression process. As the speech coder characterizes the input signal, it looks up information in a code book table and selects the code that most accurately represents the input signal. For the IS-136 TDMA system, the compression is 8:1.

As speech data compression technologies have developed, improved speech coders have become available. Hence, IS-136 can use either the original IS-54B *vector sum excited linear predictive* (VSELP) speech coder or the IS-641 *enhanced full rate* (EFR) *algebraic code excited linear predictive* (ACELP) speech coder. The EFR codec provides voice quality comparable to the landline reference *adaptive differential pulse coded modulation* (ADPCM) under normal radio channel conditions. Additionally, the EFR errored channel performance results in significant voice quality improvements.

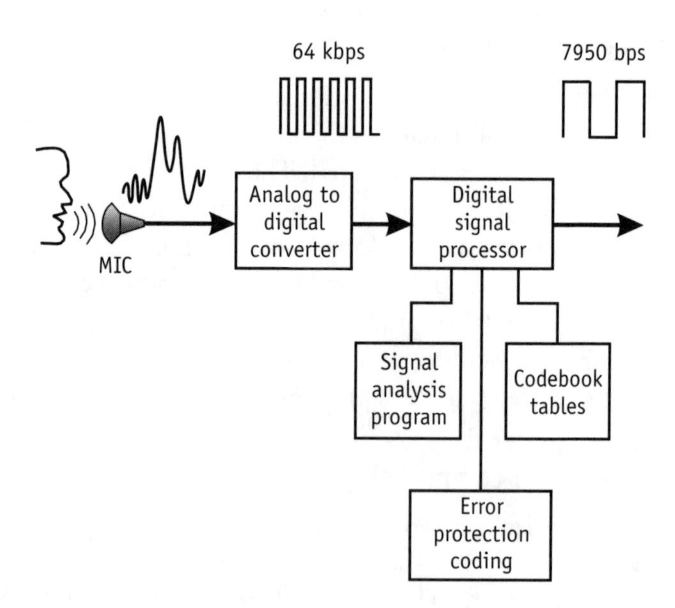

Figure 3.2 Speech coding.

3.2.3 Channel coding

After the digital speech information is compressed, control information bits are added along with error protection bits. Control messages are either time-multiplexed (simultaneously sent), or they replace (blank and burst) the speech information. Error protection bits offer a way to detect errors and to correct some errors that are introduced during radio transmission. Error protection for the IS-136 system consists of block coding and convolutional (continuous) coding. Block coding adds bits to the end of a frame (usually after several hundred bits) of information. These bits allow the receiver to determine if all the information has been received correctly or whether information should be retransmitted.

Convolutional coding involves creating unique data bits that represent the original data information combined with an error protection coding process. This information is sent simultaneously with the actual data to be transmitted. Convolutional coders are described by the relationship between the number of bits entering and leaving the coder. For example, a 1/2-rate convolutional coder generates two bits for every one that enters. The larger the relationship, the more redundancy and the better the error protection. A 1/4-rate convolutional coder has much more error protection capability than a 1/2-rate coder.

3.2.4 Modulation

The IS-136 digital radio channel uses phase modulation. Phase modulation is a process that converts digital bits into phase shifts in the radio signal. Phase modulation is a result of shifting the carrier frequency higher and lower to introduce phase changes at specific points in time.

The IS-136 digital channels use $\pi/4$ DQPSK modulation. $\pi/4$ DQPSK modulation was chosen to maintain spectral efficiency and to optimize the RF amplifier section. To create a $\pi/4$ QPSK modulated signal, typically two amplitude-modulated RF signals that are 90 degrees out of phase are combined. The digital information is represented by the signals' amplitudes; the resulting signal is at the same frequency and shifted in phase. This phenomenon allows the transfer of information, since different bit patterns input to the modulator cause specific amounts of phase shift in the output transmission. Therefore, if the received RF signal is sampled for phase transitions and amplitude at specific periods of time, it is possible to recreate the original bit pattern. The four allowed phase shifts (+45, +135, −45, and −135 degrees) represent the original binary information. The receiver looks for anticipated phase information, called a decision point.

Each two-bit stimulus input has a corresponding phase shift. The transition period between decision points is 41.15 μsec, resulting in a symbol rate of 24.3 thousand symbols per second (Ksps). Each symbol represents 2 bits, so the input data rate is 48.6 Kbps.

Digital modulation results in RF power that is distributed over a wider frequency bandwidth than AMPS FM modulation. This has resulted in a more tolerant requirement for this RF spectral density (covered in Chapter 6) than for AMPS channels. The requirement for this spectral mask was derived from adjacent channel and cochannel interference levels.

3.2.5 RF power

A major difference between digital and analog technologies is that digital requires a linear RF amplifier. A linear amplifier distorts the signal less than the class C RF amplifiers used in AMPS cellular telephones. Unfortunately, the battery-to-RF energy conversion efficiency for linear amplifiers is 30–40% compared with 40–55% for class C RF amplifiers used in AMPS phones. Linear amplifiers require more input energy to produce the same RF energy output power during transmission. Digital technologies overcome this limitation either by transmitting for shorter periods or by precisely controlling power to transmit at lower average output power.

The IS-136 system adds a new power class of mobile phone. The class IV mobile phone output power is identical to class III, but its minimum power is 12 dB lower. The lower minimum power allows systems to reduce the minimum cell site radius. Table 3.2 shows the power classification types for the IS-54B radio system.

3.2.6 Frequency allocation

IS-136 uses the existing cellular radio channels in the 850-MHz band as well as the radio channels at 1,900 MHz. In either band, the channels are

Table 3.2
IS-136 Power Classification

RF Power	Class I	Class II	Class III	Class IV	Class V–VI
Maximum power	4W	1.6W	.6W	.6W	Reserved
Average power (full rate TDMA)	1.333W	.533W	.2W	.2W	Reserved
Minimum power	6 mW	6 mW	6 mW	0.5 mW	Reserved

30-kHz wide. Figure 3.3 displays the IS-136 frequency allocation. Note that EIA 553 *analog control channels* (ACCs) and analog voice channels are

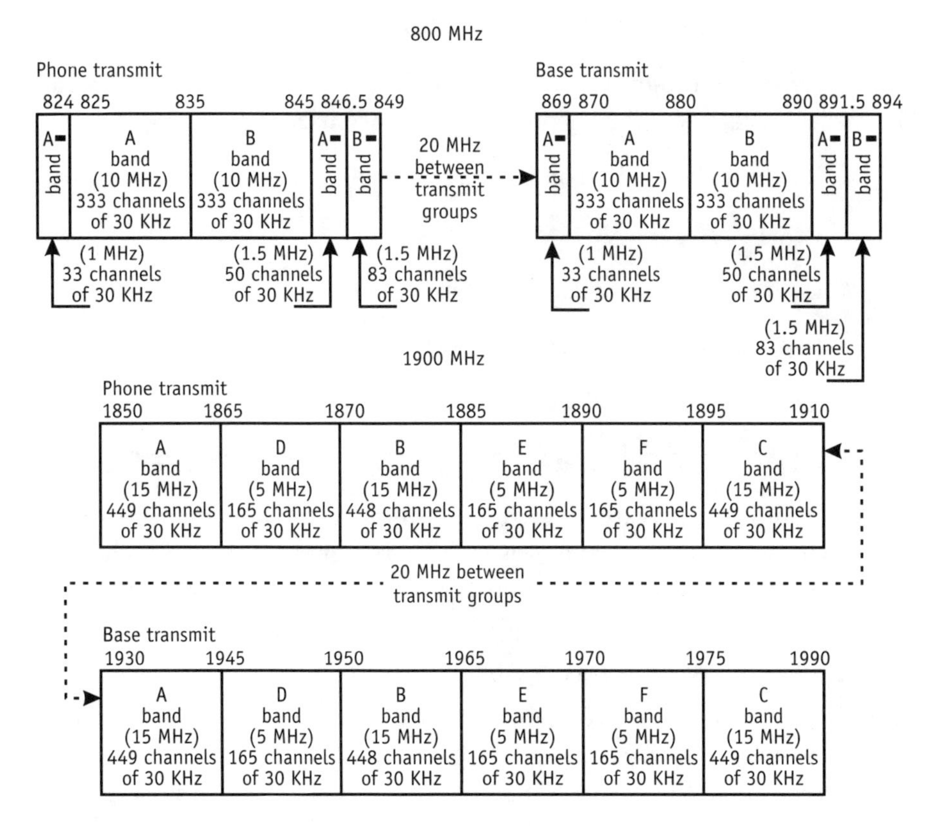

Figure 3.3 The 800-MHz and 1,900-MHz spectrum.

only available at 800 MHz. TDMA digital traffic and DCCHs are, however, available at both 800 and 1,900 MHz for seamless access to new features.

To calculate the frequency of the cellular radio channel, the following formulas, in which N represents the channel number, are used:

- For the 800-MHz band reverse channel:

  1 to 799: $0.03(N) + 825$ MHz;

  990 to 1023: $0.03(N - 1023) + 825$ MHz.

- For the 800-MHz band forward channel:

  1 to 799: $0.03(N) + 870$ MHz;

  990 to 1023: $0.03(N - 1023) + 870$ MHz.

To calculate the frequency of the PCS radio channel, the following formulas, where N is the channel number, are used:

- For the 1,900-MHz band reverse channel:

  1 to 1999: $0.03(N) + 1849.980$ MHz.

- For the 1,900 MHz band forward channel:

  1 to 1999: $0.03(N) + 1930.020$ MHz.

3.3 The digital traffic channel environment

30 kHz radio channels are divided up to support many digital users in the same bandwidth as one analog user. This requires a new radio channel structure in addition to new modulation and signaling.

3.3.1 Radio channel structure

The digital channel is frequency-duplex, meaning that the transmit and receive operations take place on different frequencies. These receive and transmit frequencies are divided into time slots that also allow *time division duplex* (TDD) operation. Forward and reverse channels are separated by 45 MHz in the 800-MHz band and by

80.04 MHz in the 1,900-MHz band. The transmit band for the 800-MHz base station is 869–894 MHz and 1,930–1,990 MHz for the 1,900-MHz base station. The mobile transmit frequency is 824–849 MHz for the 800-MHz band and 1,850–1,910 MHz for the 1,900-MHz band. The same 30-kHz channel bandwidth is the same for AMPS and IS-136 *digital traffic channels* (DTCs). Figure 3.4 shows the IS-136 radio channel structure.

To prevent a mobile phone's receive and transmit messages from interfering with each other, IS-136 systems separate transmit and receive frequencies and separate transmission and reception in time. The time separation or offset simplifies the design of transmitters and receivers. Figure 3.5 shows IS-136 system time and frequency separations between receive and transmit channels.

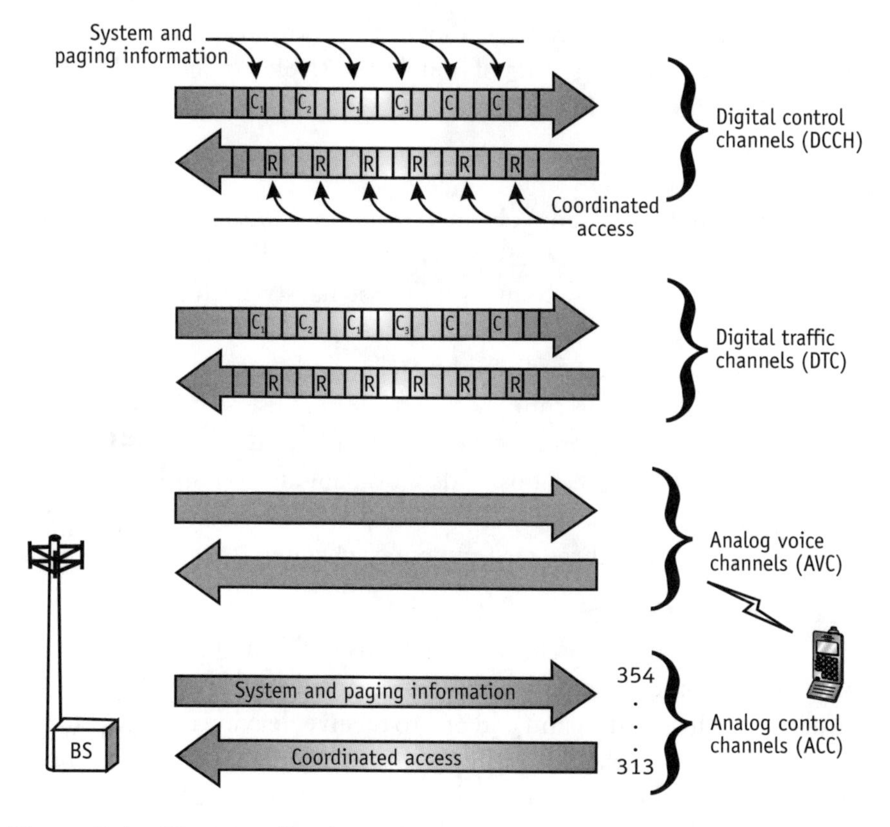

Figure 3.4 IS-136 radio channel structure.

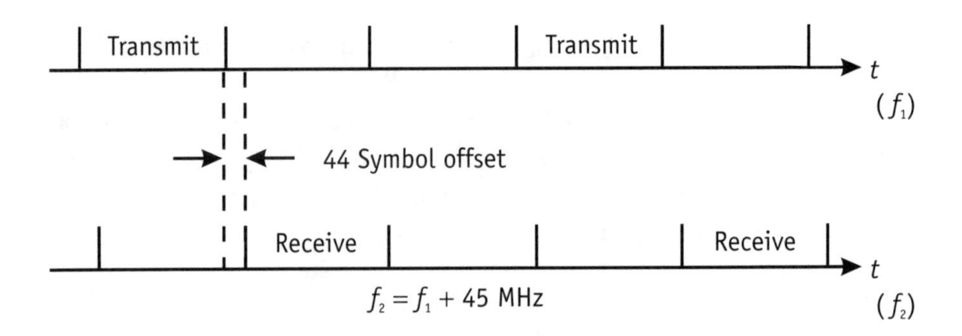

Figure 3.5 IS-136 radio channel time offset.

There are two types of channel usage available in an IS-136 system: Full-rate IS-136 systems allow three users to simultaneously share a radio channel. Half-rate IS-136 systems allow six users to share a radio channel. Other channel usage consisting of double and triple rate channels will be available for data users in the future.

3.3.1.1 Full-rate TDMA

For the full-rate IS-136 radio channel, two time slots are used for transmitting, two are used for receiving, and two are idle. The mobile phone typically uses the idle time to measure the signal strength of surrounding channels to assist in hand-off. Mobile phones transmit every third slot so that phone #1 uses slots 1 and 4, phone #2 uses slots 2 and 5, while phone #3 uses slots 3 and 6. This time sharing results in a user-available data rate of 13 Kbps. However, some of the user data is used for error detection and correction, leaving 8 Kbps of data available for compressed speech data. Figure 3.6 shows how TDMA full-duplex radio channels are divided in time to serve up to three customers per channel.

3.3.1.2 Half-rate TDMA

A radio channel's capacity can be doubled by dedicating only one slot per frame per customer, creating a half-rate channel. Half-rate channels use one of the six slots to transmit and one to receive, leaving four idle. In this case, a half-rate vocoder operating at 4 Kbps would be required to take advantage of the smaller bandwidth allocation. Figure 3.7 illustrates the half-rate TDMA channel structure.

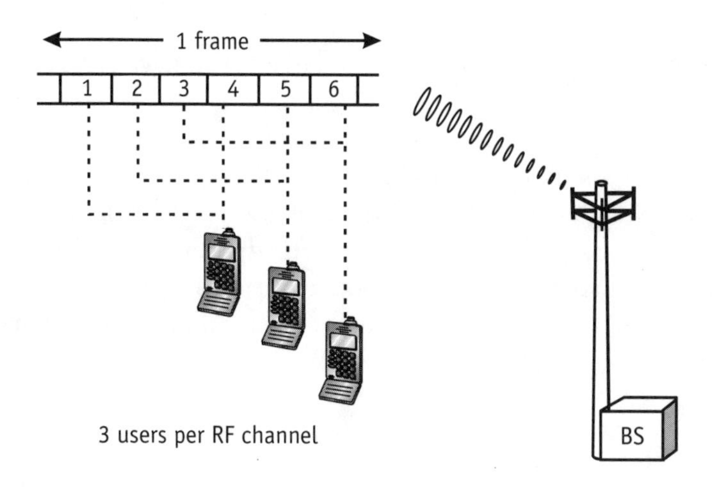

Figure 3.6 IS-136 full rate.

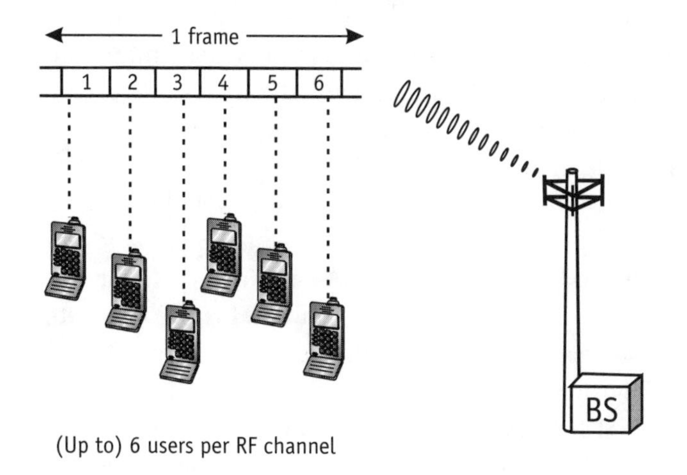

Figure 3.7 TDMA IS-136 half rate.

3.3.2 Slot structure

Each IS-136 digital channel is divided into 40 msec frames each composed of six 6.67 msec time slots. Mobile phones either transmit, receive or remain idle during a time slot. The IS-136 standard describes several slot structures that are used to transfer voice and user data. These include forward speech slot, reverse speech slot, FACCH data message slot for in band control messaging, and shortened burst slot. Each slot is composed of 324 bits (162 symbols).

Interleaving, or the continuous distribution of data bits between adjacent slots, is used to overcome the effects of burst errors due to Rayleigh fading. Diagonal interleaving is used so that the information, including errors, is distributed between adjacent slots. This distribution helps the error protection process since consecutive (burst) errors are spread, thereby enabling the error protection code to work more accurately.

3.3.2.1 Forward data slot

The forward data slot transfers voice and data traffic from the base station to the mobile phone. It contains 324 data bits, 260 of which are available to the subscriber. The initial field in the slot contains the synchronization field that identifies the slot number and provides timing information for the decoder. It is a standard pattern that may also be used for equalizer training. The equalizer adjusts the receiver to compensate for radio channel change (distortion). The SACCH field contains a set of dedicated bits for sending control information. The data fields carry the subscribers' voice and data information. The *coded digital verification color code* (CDVCC) is similar in function to SAT in analog cellular where each cell is referenced by a unique identifier. This helps the phone to distinguish between two cells that are using the same frequency. The slot format is shown in Figure 3.8.

Every IS-136 forward traffic channel slot format includes a *coded digital locator* (CDL) field that indicates a range of eight RF channels where the DCCH can be found. This field helps a mobile phone find a DCCH during an initial scan.

3.3.2.2 Reverse data slot

The reverse data slot transfers voice and data from the mobile phone to the base station. It differs from the forward data slot in that it includes guard and ramp time. During the guard time period (approximately 123 μsec), the mobile phone's transmitter is off. Guard time protects the system from bursts being received outside the allotted time slot interval due to the propagation time between the mobile phone and cell site (see

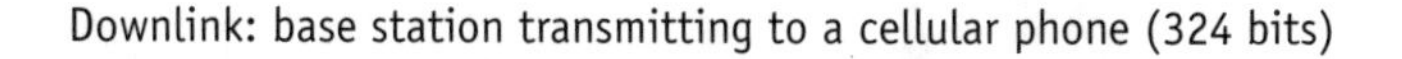

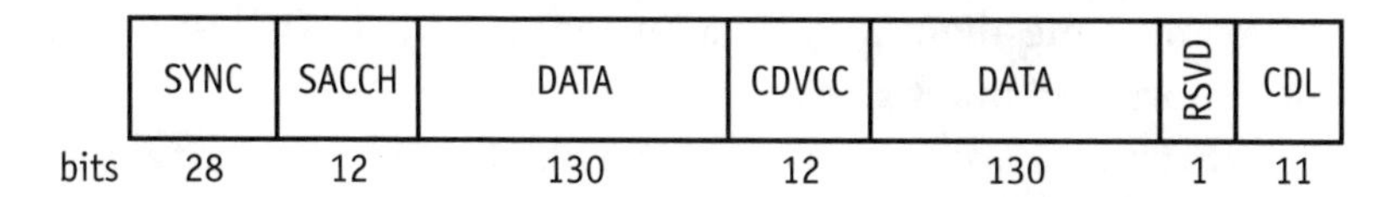

SYNC = synchronization and timing
SACCH = slow associated control channel
CDVCC = coded digital verification color code
DATA = coded speech information
CDL = coded digital locator
RSVD = reserved

Figure 3.8 Downlink TDMA burst format.

dynamic time alignment). Ramp time slowly turns on the transmitter to protect other mobile phones from interference (outside the allotted 30-kHz bandwidth) that occurs if a mobile phone turns on instantaneously. The synchronization word, CDVCC, and SACCH fields provide the same functions as described in the forward traffic channel slot. Figure 3.9 illustrates the slot format.

3.3.2.3 Shortened burst

When a mobile phone begins operating in a large diameter cell or following a hand-off between two adjacent cells of very different size, it sends shortened bursts until the appropriate timing can be established with the system. Radio link propagation time in large cells could be so long (with a round trip in excess of 500 μsec) that overlapping bursts could cause significant problems. The shortened burst allocates another guard time,

Uplink: cellular phone transmitting to a base station (324 bits)

	G	R	PREAM	SYNC	DATA	SYNC+	DATA
bits	6	6	16	28	122	24	122

G = guard time
R = ramp time
PREAM = preamble
SYNC = synchronization and timing
DATA = DCCH information
SYNC+ = additional synchronization

Figure 3.9 Uplink TDMA burst format.

thereby preventing received bursts from overlapping before a mobile phone's dynamic time alignment has adjusted. Figure 3.10 illustrates the format of a shortened burst slot.

Notice that there are several synchronization fields. Any two sync fields are separated by different amounts of time so that a base station receiver can simply detect the relative time the burst is being received in comparison to other bursts. If bursts are received out of their expected time periods, the base station can command the mobile phone to adjust its transmit time. After the shortened burst has been used to determine time alignment, the mobile phone will begin to use the standard reverse traffic channel slot structure to send user information.

3.3.2.4 FACCH data slot

When urgent control messages such as a hand-off command are sent, signaling information replaces speech information (260 data bits) in a manner similar to the blank and burst process used for control on the AMPS voice channel. The *fast associated control channel* (FACCH) message slot is identified by use of a different type of error correction coding. Initially, all slots are decoded as speech data slots, but if an FACCH message is in a speech data slot, the CRC check sum and other error detection code outputs will fail, and the message will be decoded as an FACCH data slot. This process is used so that information bits need not be dedicated to indicate whether a data slot is for speech or control. The FACCH data slot

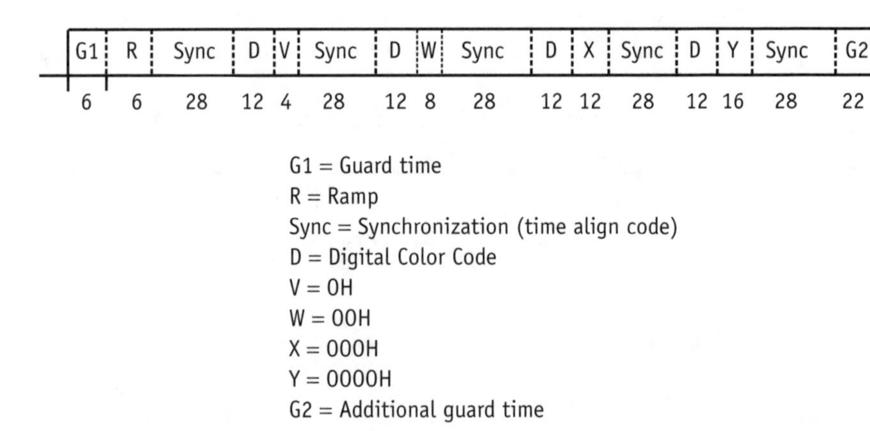

Figure 3.10 Shortened burst slot structure.

structure is identical to a speech slot, and only the data bits are FACCH data rather than digitally coded speech.

3.3.3 Traffic channel signaling

The DTC offers several methods of transferring control information. These can be divided into in-band and out-of-band signaling. In-band signaling replaces voice data, and out-of-band is sent simultaneously with the voice data. In-band signaling for the DTC includes FACCH. Out-of-band signaling is called *slow associated control channel* (SACCH).

3.3.3.1 Slow associated control channel (SACCH)

SACCH is a continuous data stream of signaling information sent beside speech data (out-of-band signaling). SACCH messages are sent using dedicated bits in each slot, so SACCH messages do not affect speech transmission. However, the transmission rate for SACCH messages is slow. For rapid message delivery, control information is sent via the FACCH channel. The SACCH and FACCH system was designed to maximize the number of bits devoted to speech and optimize the number of bits devoted to continuous signaling.

The SACCH is allocated 12 bits per slot. A message is composed over 12 slots, resulting in a gross rate of 600 bps. The data is 1/2-rate convolutionally coded, reducing data transmission to 300 bps, including control flags and CRC. Figure 3.11 shows the SACCH signaling process.

3.3.3.2 Fast associated control channel (FACCH)

FACCH control messages replace speech data with signal messages (in-band signaling). FACCH data is error-protected by a 1/4-rate convolutional coder, which increases the error protection since control messages are often sent in poor radio conditions (e.g., hand-off). FACCH messages use the entire 260 data bits of the burst, providing a gross data rate of 13 Kbps. However, the 1/4-rate convolutional coding reduces the data transmission rate to 3,250 bps. Speech quality can be degraded as more and more speech frames are replaced with signaling information, and infrastructure manufacturers are careful not to demand too many consecutive FACCH transactions during normal operation. Figure 3.12 shows FACCH signaling.

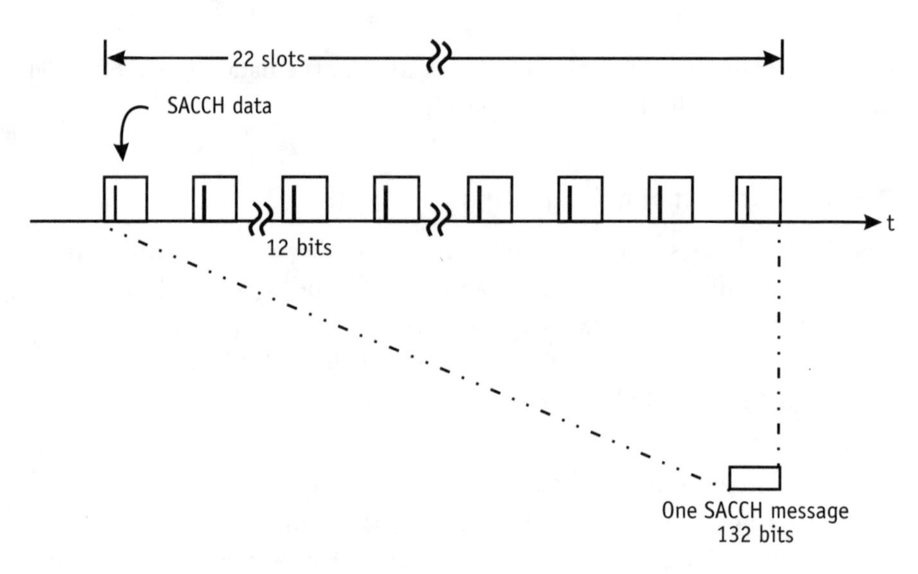

Figure 3.11 SACCH signaling.

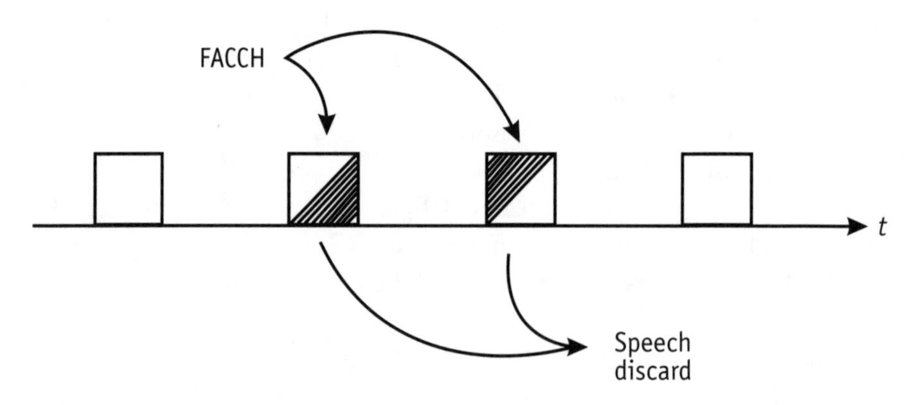

Figure 3.12 FACCH signaling.

3.3.3.3 Digital verification color code (DVCC)

Each cell site in a cellular system (or localized region of the system) has its own unique DVCC code. This is used by phones to detect interference from neighboring cell sites while decoding system information. A unique DVCC for each cell site ensures that the correct mobile phone is communicating with the proper base station, since frequencies are reused in most cellular systems.

The base station sends the DVCC and adds four parity bits to the eight-bit DVCC code. This is the CDVCC. The value 00h is not used as a valid CDVCC, leaving 255 unique codes.

3.3.3.4 Dual-tone multifrequency signaling

Digital voice coders are not designed to handle nonspeech audio signal such as DTMF tones. As a result, they can change the amplitude relationship between one tone component and another within the DTMF signal.

To avoid speech coder DTMF distortion, a DTMF on message can command the base station or MSC to create DTMF tones. Figure 3.13 shows a user pressing key number 2 (step 1) to create an FACCH message (step 2) that indicates that digit #2 has been pressed. The receiver in the base station decodes this message (step 3) and commands a DTMF generator to create a number 2 DTMF touch tone (step 4). When the user releases the #2 key, an FACCH message is created, indicating that the #2 key has been released and that the DTMF touch tone is stopped. In addition to the PRESS and RELEASE messages, the mobile station can also send a message that contains one or more DTMF digits. This message is used for "speed dial" service or similar purposes. The duration of each digit is preset according to the industry standards.

3.3.4 Dynamic time alignment

Dynamic time alignment is a technique that allows the base station to receive digital mobile phones' transmit bursts in an exact time slot. Time alignment keeps different digital subscribers' transmit bursts from colliding or overlapping. Dynamic time alignment is necessary, because

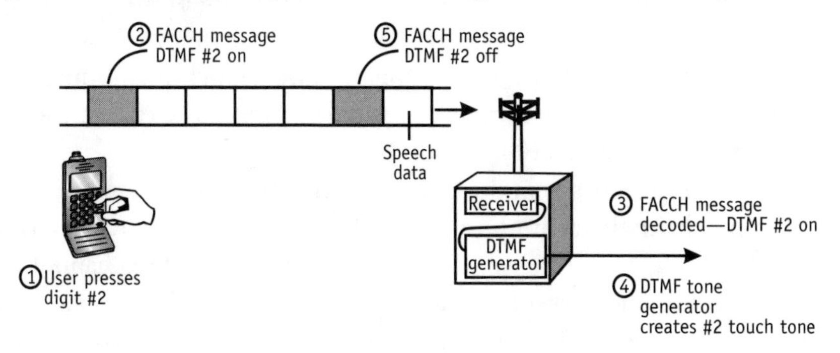

Figure 3.13 DTMF signaling.

subscribers are moving or scattered within the cell coverage, and the arrival time of signals at the base station depends on their changing distance from the base station. The greater the distance, the more delay in the signal's arrival time. Dynamic time alignment adjusts for differences in the signal travel time according to each mobile phone's distance from the base station.

The base station adjusts for the delay by commanding mobile phones to alter their relative transmit times based on their distance from the base station. The base station calculates the required offset from the mobile phone's initial transmission of a shortened burst in its designated time slot (necessary only in large diameter cells where propagation time is long). To account for the combined receive and transmit delays, the required timing offset is twice the path delay. The mobile phone uses a received burst to determine when its burst transmission should start. The mobile phone's default delay between receive and transmit slots is 44 symbols, which can be reduced in 1/2-symbol increments to 15 symbols. Figure 3.14 shows the need for time alignment.

3.3.5 Mobile assisted hand-off (MAHO)

MAHO is a system in which the mobile phone assists the MSC with hand-off decisions by sending radio channel quality information back to the system. In existing analog systems, hand-off decisions are based only on measurements of mobile phones' signal strength made by receivers at the base station. IS-136 systems use two types of radio channel quality information: signal strength of multiple neighbor channels and an estimated bit error rate of the current channel. The bit error rate is estimated using the result of the forward error correction codes for speech data and call processing messages. Having the mobile phone report quality information also allows for measurements of the downlink quality that are not possible from the base station.

The system sends the mobile phone a MAHO message containing a list of radio channels from up to 24 neighbor cells. During its idle time slots, the mobile phone measures the signal strength of the channels on the list including the current operating channel. The mobile phone averages the signal strength measurements over a second, then continuously sends MAHO channel strength reports back to the base station every

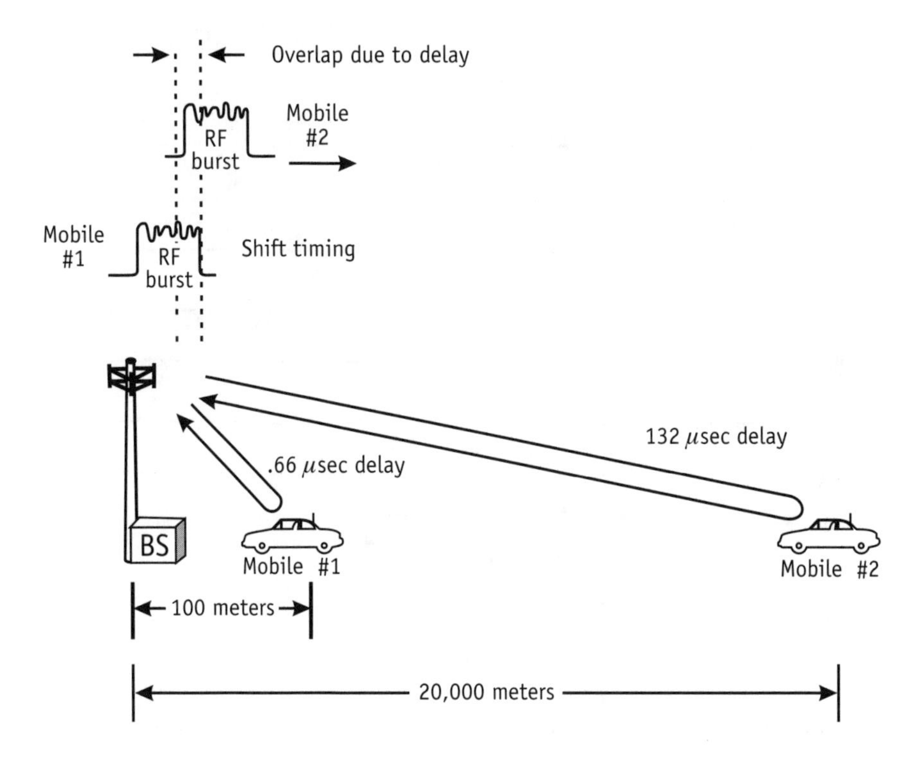

Figure 3.14 Dynamic time alignment.

second. The system combines the MAHO measurements with its own information to determine which radio channel will offer the best quality, and it initiates hand-off to the best channel when required. Figure 3.15 illustrates the MAHO process.

3.4 The digital control channel environment

The digital control channel is used to convey call setup information and to provide the platform for enhanced services in an IS-136 system. This requires signaling, messaging, and an advanced radio channel structure.

3.4.1 DCCH basic operation

The DCCH is introduced into the wireless system by defining one DCCH slot pair on a frequency that contains existing DTCs. These DCCHs are

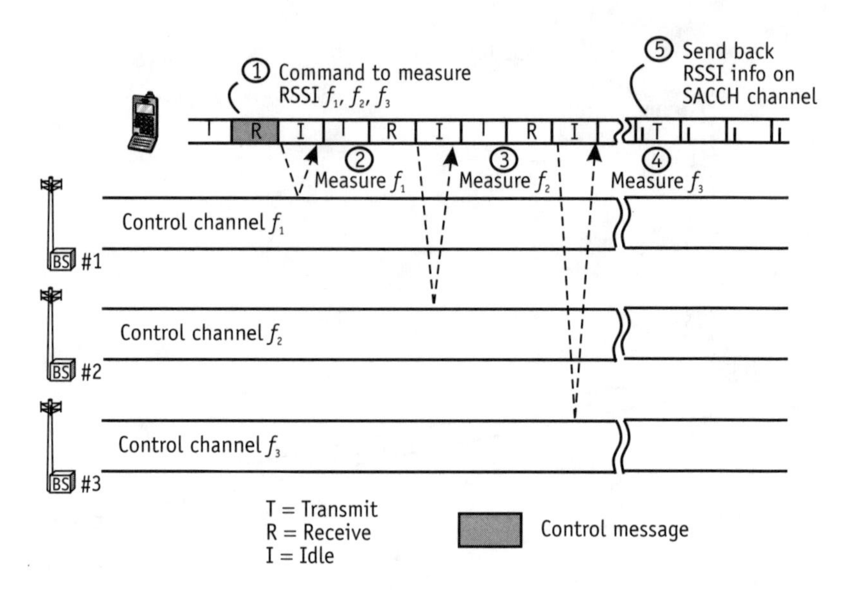

Figure 3.15 Mobile assisted hand-off.

not restricted to the 21 channels used by the ACCs and can be anywhere in the 800- or 1,900-MHz bands. DCCH capable phones will monitor (camp on) a DCCH instead of an ACC in each sector of a system that supports IS-136 services. DCCH-capable phones will scan for this channel, gain synchronization, and begin to decode the information provided over a broadcast control channel on the DCCH. The DCCH will serve as the phone's control channel until the phone finds another cell that is more appropriate. Figure 3.16 shows the 1,4 slot pair used for a DCCH.

The DCCH-capable phone will receive pages, send originations, and communicate with the system on the DCCH. After receiving a page or performing a call origination, a traffic channel is then designated for the duration of the call. That is, after call setup, the cellular phone will retune to the DTC or analog voice channel and the conversation will take place. At call completion, the phone will return to the DCCH instead of the ACC. If no DCCH is available, the phone can obtain service on an ACC.

3.4.2 DCCH burst format

The TDMA DTCs and the DCCHs both use the standard TDMA frame structure, which uses three slot pairs to allow three digital conversations

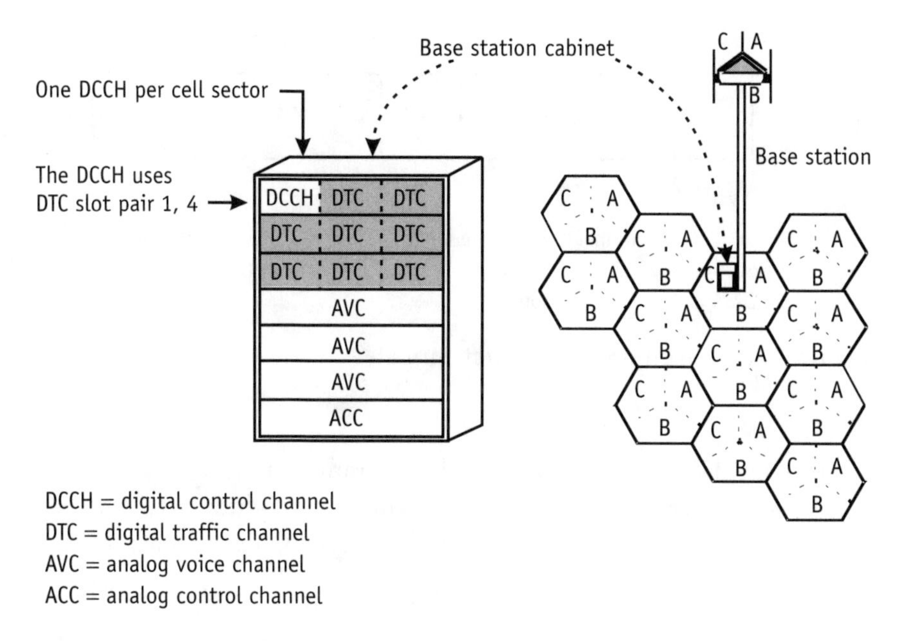

Figure 3.16 IS-136 DCCH operation.

to be carried on one frequency. One of these slot pairs is used for a full-rate DCCH in each sector of a cell. Generally only one slot pair is required for a DCCH in each cell sector to serve as the control link for the call control information. This means that all the control signaling is performed in the same bandwidth as one DTC.

3.4.2.1 Forward control channel slot (downlink burst)

The IS-136 specification defines a downlink burst format for the DCCH. Figure 3.17 shows the IS-136 DCCH burst format.

The downlink fields of the DCCH burst differ from those of the DTC burst in the following ways:

- The SACCH field is replaced by the *shared channel feedback* (SCF) field. This field is a collection of flags used as a method of control and acknowledgment of information sent from the phone to the base station.
- The voice field data is replaced by DCCH data.

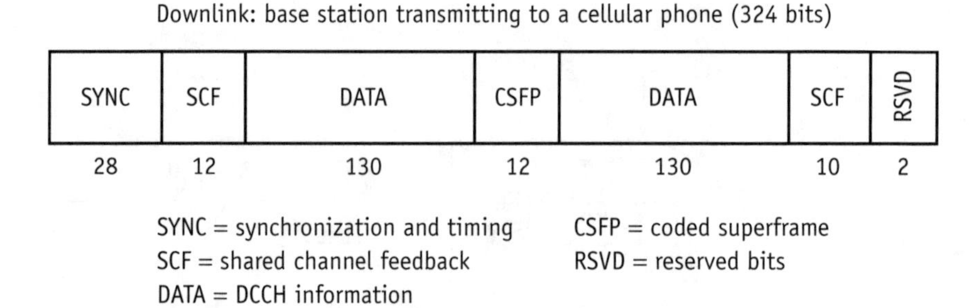

Figure 3.17 DCCH downlink burst format.

- The CDVCC field is replaced by a frame-counting field called the *coded superframe phase* (CSFP). This field indicates to the phone which frame in the superframe is currently being transmitted.
- The RSVD field is replaced by the remaining bits of the SCF field.

Because the DCCH burst copies the basic TDMA burst structure used for the traffic channels, a DCCH will be perceived as a normal traffic channel by a non-DCCH-capable phone.

3.4.2.2 Reverse control channel slots (normal and abbreviated)

The base station has to treat each uplink DCCH burst of data from a phone as a unique transmission and has to achieve time alignment and bit synchronization on each DCCH burst. A *preamble* (PREAM) sequence and an additional *synchronization* (SYNC+) word is placed in each uplink packet to enable the base station to lock onto single bursts of data from phones and decode the uplink information.

The time alignment methods used for the traffic channels, which rely on a symbol-by-symbol advance or retreat of consecutive transmissions for the phone, are not possible on the uplink DCCH because of its single-burst nature. Therefore, to prevent a difference in transmit times from causing a misalignment of received bursts at the base station, there are two uplink burst lengths—a normal burst for small cells and an abbreviated burst for large cells where time alignment might be an issue. The appropriate burst length to use is set by the system operator and is announced to a phone in the broadcast information for each cell.

An abbreviated burst is used to correct the relative time offset from near and distant cellular phones within a large cell. Using the shorter burst length in large cells reduces the probability of the burst overlapping a frame when it is received at the base station. Figures 3.18 and 3.19 show the normal and abbreviated DCCH slot formats for the normal control channel uplinks slot structure.

The PREAM is used for timing. It is also used by the base station to set the receiver amplifier to avoid signal distortion. The SYNC field is a known pattern that allows the base station to find the start of the incoming TDMA burst. The data field, which is the payload, is divided into a two-part field. The SYNC+ is another fixed bit pattern that provides additional synchronization information for the base station.

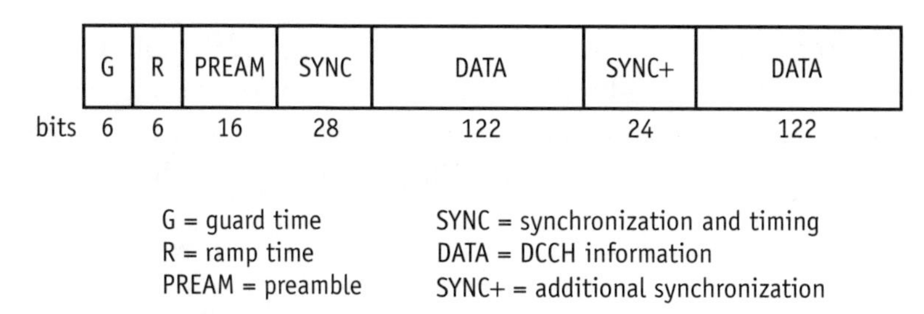

Figure 3.18 DCCH normal uplink burst format.

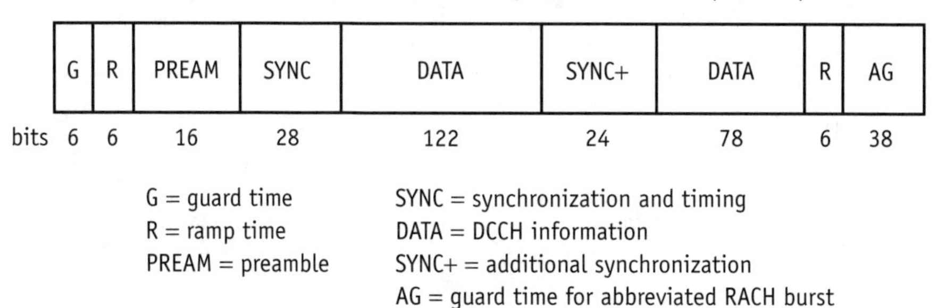

Figure 3.19 DCCH abbreviated uplink burst format.

3.4.3 Superframe and hyperframe formats

Superframe and hyperframe formats are used to multiplex logical groups of information together and to provide a known repeatable sequence on the air interface. This enables a phone to retrieve information quickly and to develop a sleep mode in which the phone only needs to wake up at predefined instances to receive messaging.

3.4.3.1 Superframes

A superframe is made up of sixteen sequential 40-ms TDMA frames equivalent to 32 consecutive TDMA blocks at full rate. Only slots one and four are used to carry DCCH information; this creates a sequence of 32 DCCH carrying bursts spread through 96 TDMA bursts. Each DCCH burst in the superframe is designated for either broadcast, paging, *short message service* (SMS) messaging, or access response information. The superframe structure shown in Figure 3.20 is continuously repeated on the DCCH channel.

The *broadcast channel* (BCCH) shown in Figure 3.21 is split between a *fast BCCH* (F-BCCH) used for mandatory information with a guaranteed data throughput, and an *extended BCCH* (E-BCCH) used for additional information of a less critical nature. These BCCHs are indicated respectively as F and E in Figure 3.21.

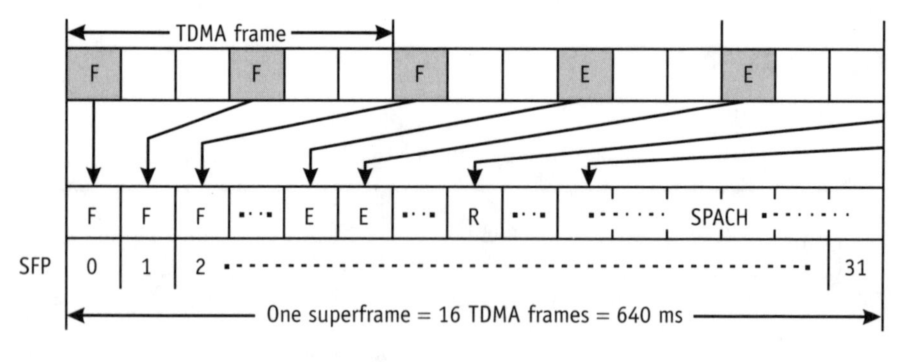

Figure 3.20 Superframe structure.

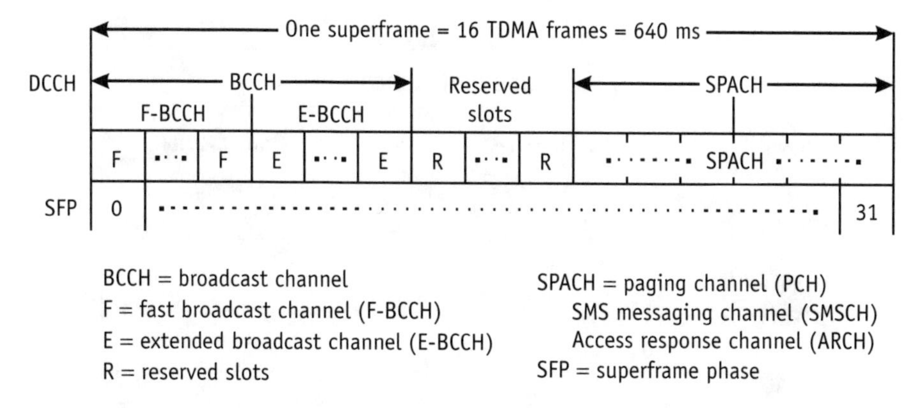

Figure 3.21 Logical channels multiplexed onto the superframe.

The SPACH channel shown in Figure 3.21 is comprised of the *paging channel* (PCH), the *SMS messaging channel* (SMSCH), and the *access response channel* (ARCH). These logical channels are described in more detail later.

The superframe is created by multiplexing the broadcast and other logical channels in a repetitive, ordered sequence onto a physical DCCH burst on the downlink (base station to phone). Figure 3.21 shows the channels multiplexed onto the superframe.

Table 3.3 defines the number of slots that can be supported for each logical channel. This allows the superframe to be tuned to meet the needs of specific environments. For example, an operator may use very few DCCH neighbors and would prefer to provide more bandwidth for paging messages. The superframe structure is flexible and is broadcast to phones when a DCCH is first acquired. The *superframe phase* (SFP) increments every superframe slot, starting at 0 on the first F-BCCH slot and counting modulo 32.

Note: All of the time slots on the uplink (phone to base station) are used for system access by the phone on the RACH (random access) logical channel. This is described in Section 3.5.3.

A phone learns the structure of the superframe from information transmitted in the F-BCCH. When a cellular phone first finds a DCCH, it must determine the slot in the superframe that is actually being monitored. This is achieved by monitoring the CSFP field in the DCCH burst. When the first slot of the superframe is encountered, the phone can

Table 3.3
Superframe Slot Allocations

Slots	Full-Rate DCCH		Half-Rate DCCH	
	MIN	MAX	MIN	MAX
F-BCCH (F)	3	10	3	10
E-BCCH (E)	1	8	1	8
Reserved (R)	0	7	0	7
SPACH	1	32 – (F + E + R)	1	16 – (F + E + R)

decode the F-BCCH information (since it is always transmitted first) and ascertain the structure of the superframe. The phone then knows the slot usage for the rest of the superframe and can decode the E-BCCH and SPACH information.

3.4.3.2 Hyperframes

A hyperframe is made up of a primary and secondary superframe as shown in Figure 3.22. The hyperframe length consists of 192 TDMA bursts. Sixty-four of these (or every third) are used at full rate for DCCH

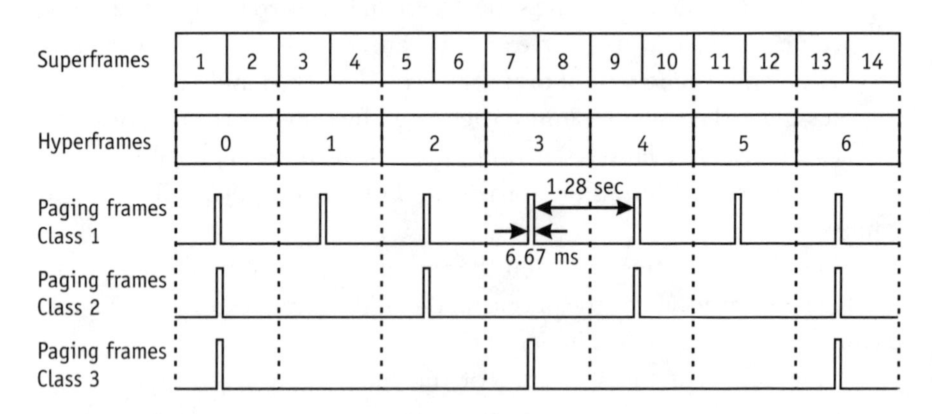

Figure 3.22 Hyperfames showing paging frame classes.

information to maintain the same burst usage as a TDMA traffic channel. By repeating the DCCH information in each hyperframe, the hyperframe structure allows a cellular phone to read broadcast-channel information on adjacent, nonsynchronized DCCHs and scan for another DCCH on a different frequency, without missing a page on its own DCCH.

To enhance the readability of pages under severe radio conditions, all pages are repeated at the corresponding time slot in the second superframe. By having a specified page repeat, if a phone cannot correctly decode the first assigned PCH slot in the first half of the hyperframe, it can read the corresponding PCH slot in the secondary superframe (640 ms later). In addition, a level of redundancy is built into the DCCH so that burst errors—that is, a cluster of closely spaced bits that are incorrect—will not affect both superframes.

3.4.3.3 Paging classes

Paging frame classes determine the time period (in hyperframes) in which the system can potentially page a phone. Since each phone is assigned a particular paging slot on the DCCH superframe, the phone knows precisely when it is likely to get paged and need only monitor that particular paging slot. This is the basis of the sleep mode described in Section 3.4.3.4.

There are eight paging frame classes supported, ranging from one to 96 hyperframes, which represent 1.28 to 123 seconds between paging opportunities. The eight classes comprise 1, 2, 3, 6, 12, 24, 48, and 96 hyperframes duration. Voice services will use lower paging-frame classes because the higher paging-frame classes, reserved for nonvoice applications, will introduce unacceptable call-setup delay.

3.4.3.4 Sleep mode

Since the superframe structure is known, the phone only needs to monitor downlink DCCH information on its predetermined paging slot. This will provide extended periods of time in which the phone can power down some of its circuitry and sleep between paging opportunities, thus saving battery standby time.

While a phone is in idle mode and waiting for pages (for either a voice call or teleservice message), indicators in the time slots used for paging inform the phone about broadcast information changes. Thus, as long as the broadcast information does not change, the phone only has to wake up and read its paging slot and perform channel measurements. This provides for an efficient sleep mode in conjunction with a fast response to broadcast updates (for instance, during cell changes).

Since phones spend most of their time waiting for a page, the PCH structure directly affects their *sleep* time. The IS-136 PCH was designed to maximize the sleep time available to a phone, thereby increasing the battery standby time.

3.5 Logical channels

Logical channels have been created in IS-136 to organize the information flowing across the air interface as shown in Figure 3.23.

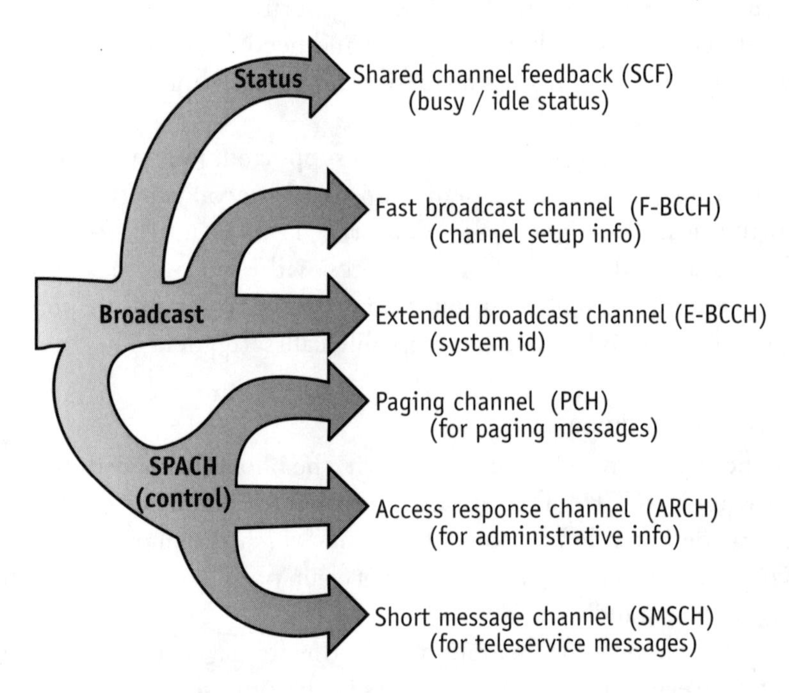

Figure 3.23 Logical channels.

3.5.1 Broadcast channel (BCCH)

The BCCH continuously provides information about the system configuration and the rules that a cellular phone must follow at system access. The BCCH logical channels are multiplexed onto the physical DCCH downlink channel as part of the superframe. The BCCH information includes SID, neighbor lists of other DCCHs for DCCH reselection, DCCH frame structure, and other system information.

3.5.1.1 Fast BCCH and extended BCCH

The BCCH channel is divided into F-BCCH and E-BCCH. Another dedicated channel, the *short messaging service BCCH* (S-BCCH), is being created to provide a broadcast messaging service.

The first timeslot of a superframe is always an F-BCCH slot. The F-BCCH information is sent in its entirety once every superframe (640 ms), whereas a complete set of E-BCCH information might span several superframes. The F-BCCH channel is used for mandatory, time-critical system information requiring a fixed repetition cycle. The information sent on this channel relates to SID and parameters needed by a cellular phone in determining the following:

- Structure of the superframe;
- System on which the phone is camped;
- Registration parameters;
- Access parameters.

The E-BCCH channel is used for additional system information that is less time-critical (in terms of a phone needing to camp) and that does not require a guaranteed rate (for example, neighbor cell lists).

3.5.2 SMS, paging, and access channel (SPACH)

The *SMS, paging, and access channel* (SPACH) provides mobile phones with paging and system access parameter information. The SPACH channel is divided into a paging channel (PCH), an access response channel (ARCH), a short message service (SMS), and a point-to-point messaging channel (SMSCH).

3.5.2.1 Paging channel (PCH)

The PCH is used to transfer call setup pages to the phone. A cellular phone is allocated to a particular paging slot (part of the SPACH) in a superframe according to its *mobile station identity* (MSID). This strategy will always place a specific phone in the same paging slot on any DCCH with the same structure. In addition, this method minimizes paging congestion across the air interface, since the paging slots for all the phones camped on any particular DCCH will be randomly distributed across the total number of available SPACH slots.

3.5.2.2 Access response channel (ARCH)

The ARCH is used to send system responses (such as channel assignment commands) and administrative information from the system to the phone.

3.5.2.3 SMS channel (SMSCH)

The SMSCH is used to transfer point-to-point teleservice data to and from the mobile phone. This data can belong to a CMT text message, OAA NAM data, *over-the-air programming* (OAP) intelligent roaming information, or a *general UDP teleservice* (GUTS) packet.

3.5.3 Random access channel (RACH)

The *random access channel* (RACH) is a shared channel resource used by all DCCH-capable mobile phones when they attempt to access the system. The RACH is the only logical channel on the uplink of the DCCH. Uplink RACH messages could be, for example, a phone's response to an authentication request or a MS acknowledgment to a short message delivery.

The RACH supports both contention-based (random) and reservation-based (scheduled) accesses from phones. Contention-based access means that more than one phone may attempt an access on the same channel at once and the accesses may collide. If this occurs, the phones that did not manage to gain access will enter a retry state. Reservation-based access can be used during sequential transmissions from a particular phone where the system announces to all phones which one is allowed access at a certain point in time. This reserves the RACH channel for the duration of the uplink message and stops interruptions by other phones.

The uplink RACH mode is made possible by introducing SCF in the downlink. By reading the SCF information sent in every downlink burst, the phones will know the status of the next corresponding uplink burst, that is, whether it is idle, busy, or reserved.

The full-rate DCCH is defined to consist of six RACH subchannels or TDMA blocks to allow for base station and phone processing as shown in Figure 3.24. The phone monitors the downlink subchannel (in this example, subchannel 1) and waits for the SCF flag to show idle (described in Section 3.5.4). When this happens, the phone knows that the next uplink subchannel 1 will be available, and the phone transmits the first burst in that subchannel. The phone then waits for the next downlink subchannel 1 to find out if the uplink transmission was successful and whether it should send the next burst. This continues until the whole message has been sent.

This activity is only occurring on one of the subchannels. Other phones could be accessing the system simultaneously on subchannels 2 through 6 on the same DCCH.

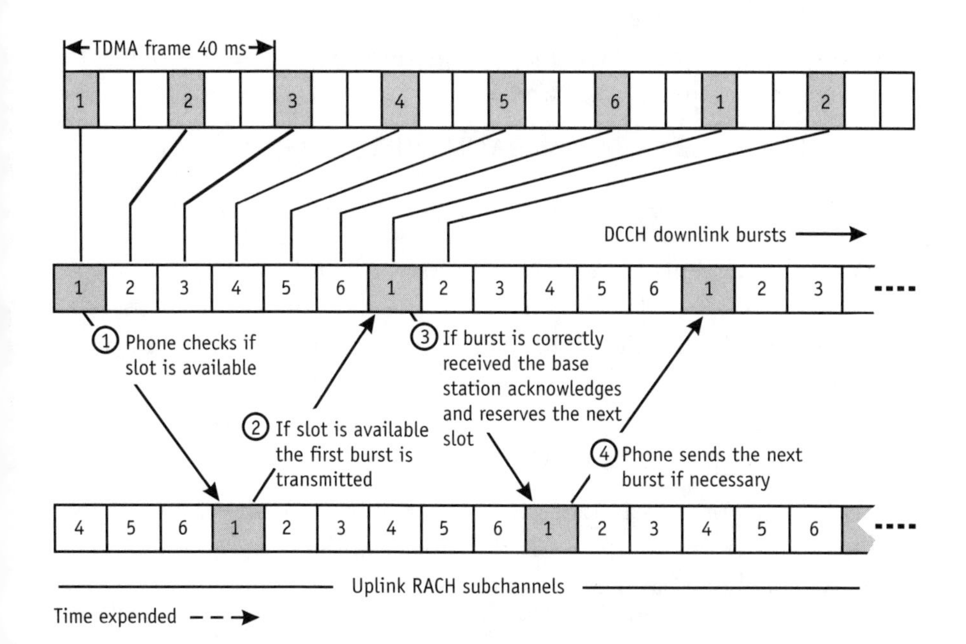

Figure 3.24 IS-136 system access on RACH channel.

3.5.4 Shared channel feedback (SCF)

The information sent on the uplink DCCH (from a phone to the base station) is controlled and acknowledged by the SCF flags located in the downlink.

The SCF downlink information provides a real-time indication of the status of every uplink timeslot and informs all phones of the usage at any particular time. In this way, the SCF provides a collision-prevention mechanism for the uplink RACH channel. This combined uplink and downlink flow of information serves to enhance the throughput capacity of the RACH. In addition, the SCF flags provide error correction for the uplink RACH channel by indicating whether or not any given burst of an access attempt has been successfully received by the system.

The SCF field indicates the status of the RACH using the *busy-reserved-idle* (BRI), *partial echo* (PE), and *received/not received* (R/N) indicator fields, which are discussed in Sections 3.5.4.1–3.5.4.3.

3.5.4.1 Busy-reserved-idle field

The base station signals the availability of the RACH by setting one of the following bits of the SCF flags.

- B: Busy, or not accepting RACH traffic;
- R: Reserved, or currently being used by a phone;
- I: Idle, or phone access available on request.

Subsequently, a phone will be able to determine if it can begin an uplink transmission.

3.5.4.2 Partial echo field

This SCF field contains a PE derived from a phone's MSID, which echoes back part of the phone ID on the downlink. In this way, all phones will know which phone has use of the RACH.

In the case of a collision during a contention-based access, the PE field is also used to indicate which phone was granted access to the system.

3.5.4.3 Received/not received field

The R/N field bits are used to indicate whether the uplink message was received correctly. This field provides the *automatic retransmission request* (ARQ) error-correction function on the uplink RACH.

3.6 Layered structure

The IS-136 air interface is structured in different layers, each with specific purposes. This conceptual split makes it easier to understand the interactions between the base station and phone across the air interface. Four layers can be identified in IS-136:

1. A physical layer (layer 1), dealing with the radio interface, bursts, slots, frames, and superframes;
2. A data link layer (layer 2) that handles the data packaging, error correction, and message transport;
3. A message layer (layer 3) that creates and handles messages sent and received across the air;
4. Upper application layers, which represent the teleservice currently being used, such as CMT, OATS, or GUTS.

The IS-136 layered structure is shown in Figure 3.25. The structure of IS-136 simplifies introduction of future services using the same DCCH platform because the lower layers in the air interface protocol (e.g., the radio interface, data management, and messages) remain unchanged.

Figure 3.26 shows how one layer-3 message is mapped into several layer-2 frames and how a layer-2 time frame is mapped onto a time slot. The time slot is further mapped onto a DCCH channel. Figure 3.26 illustrates how information is passed from layer to layer down through the stack until a TDMA burst is created, ready for transmission. At the receiving end, information is stripped off as needed as the message is passed up to the application.

The layer-3 message shown in Figure 3.26 can be an uplink registration, a downlink SMS, a page response, or a broadcast message. The length of a layer-3 message is determined by a layer-3 length indicator, which is carried as part of the layer-3 header. The length of a layer-2

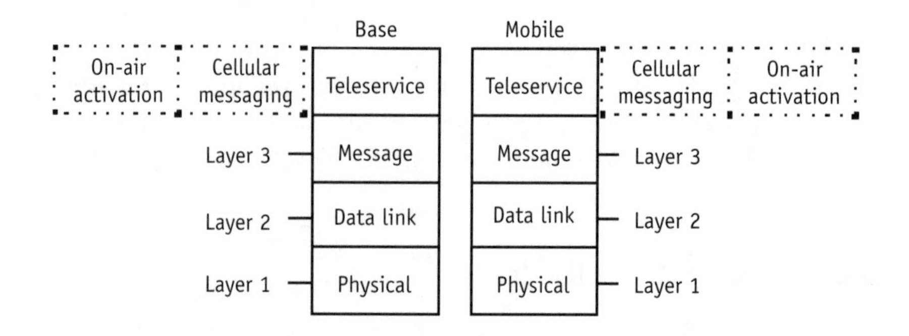

Figure 3.25 IS-136 air interface model.

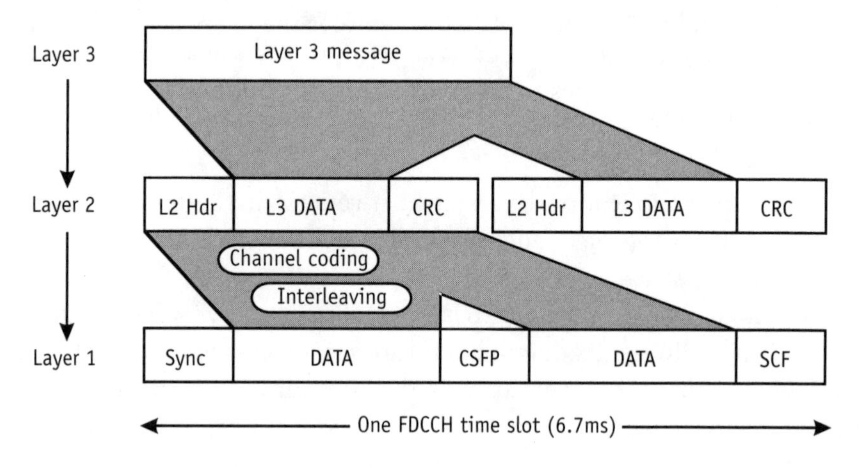

Figure 3.26 Layered 3.2.1 mapping.

frame is fixed, determined by the specific logical channel. *Cyclic redundancy check* (CRC) tail bits are added to the layer-2 frames before channel encoding.

3.6.1 Physical layer (layer 1)

The primary function of the physical layer is to transport (deliver) layer-2 information on TDMA slots, which comprise the fundamental transmission units sent over the wireless physical media. The burst structure, TDMA framing, superframes, hyperframes, and radio interface are all part of the physical layer.

3.6.2 Data link (layer 2)

Layer 2 is considered the data link layer. It provides framing and support for higher layer messages and attempts to ensure error-free transfer of layer-3 messages across the air interface. Layer-2 frames also contain overhead information for layer-2 protocol operation, handle retransmission protocols, and perform segmentation and assembly of layer-3 messages.

Each layer-2 frame is mapped into a single physical layer slot with the addition of error coding (CRC error protection) and overhead information (header information describing the type of logical channel being used). In addition, layer-2 frames allow both acknowledged and unacknowledged ARQ modes to be invoked for certain messages.

Because of the general point-to-point nature of messages sent on the SPACH logical channels, layer 2 is designed to carry the MSID to indicate the phone identity of the message recipient. The specified layer-2 protocol allows for up to five distinct MSIDs to be included within a single layer-2 frame. This effectively allows for up to five phones to be paged in one burst if 24-bit MSIDs are used. The MSID used to identify a phone can be either the MIN, the *temporary MSID* (TMSI), or the *international MSID* (IMSI), depending on the system capability (see "Identity Structures" in Section 3.8).

3.6.3 Transport (layer 3)

Layer-3 messages include information for, among others, the following:

- Registration;
- Paging;
- DCCH structure;
- Call release information;
- *Relay data* (R-data) transport for teleservices;

- Relay of higher layer information for future applications;
- Proprietary signaling.

Layer-3 messages are put into layer-2 packets that indicate the type of layer-3 information, the message length, the cellular phone to which the message is intended, and other administrative information. A layer-3 message is parsed into as many layer-2 frames as needed and packed as tightly as possible (message concatenation) to achieve throughput capacity.

3.6.4 Application layers

Special layer-3 data packets (R-data packets) are reserved for the transport of higher layer application information. These R-data packets are identical in structure, regardless of the application they are supporting.

By using the R-data packets, the air interface becomes transparent for future applications using the same transport mechanism, and the time to market for those features can be greatly reduced.

CMT is one of the applications supported by the IS-136 protocol. Messages from this application, which contain information regarding alphanumeric paging, display, and delivery options, are transferred across the air interface in R-data packets.

3.6.5 Error procedures

Owing to the harsh radio environment—error detection and correction methods and procedures are incorporated into the digital control channel to ensure robust operation and performance.

3.6.5.1 Interleaving

The interleaving process takes blocks of bits in a burst and mixes them together so that any burst errors introduced by the radio path are spread through the resulting block of data, thereby making the errors easier to correct. The receiving end reverses the procedure, deinterleaving the data and retrieving the original bit sequence.

Intraburst interleaving is the only type of interleaving performed. Interburst interleaving across different TDMA bursts as used on traffic channels is not used on the DCCH, since it would significantly decrease sleep mode efficiency and eliminate the single burst nature of the RACH and PCH.

3.6.5.2 Convolutional coding

Rate 1/2 convolutional coding is used for all downlink logical channels. Layer-1 fields, such as the SFP in the TDMA burst, are encoded using a shortened (15,11) Hamming code. This is the same correction method used for the CDVCC and CDL on traffic channels.

3.6.5.3 Automatic repeat request (ARQ)

ARQ operation is supported for the ARCH and DTC to provide enhanced data protection. ARQ is an acknowledgment process in which consecutive blocks of a message are grouped together and a results burst is returned. The sending end can then retransmit blocks that were received incorrectly.

The ARCH provides a layer-2 selective-repeat ARQ mechanism whereby ARQ MODE packets can be used to send layer-3 information with an increased level of data integrity. Each ARQ MODE packet contains a *frame number* (FRNO) and a *transaction identifier* (TID) field to identify segments for retransmission.

3.6.5.4 Shared channel feedback

The information sent across the uplink physical layer is controlled and acknowledged by the SCF field located in the downlink. This field consists of R/N flags to acknowledge the success of the last uplink transmission. The phone can retransmit information that was incorrectly received by the base station.

3.7 Digital control channel operations

Phones that use IS-136 TDMA systems must perform ancillary tasks in order to find a suitable DCCH, inform the system of their presence (registration), make calls, and send and receive teleservice messages. Phones must also be authorized and validated (in the authentication process) prior to accessing a system.

3.7.1 System and control channel selection

Unlike ACCs, which are generally confined to channels 313 to 333, a DCCH can be placed anywhere in the 800- or 1,900-MHz spectrum. This means that a DCCH-capable phone will need to execute a more complex scanning procedure to find a DCCH.

3.7.1.1 Intelligent roaming

IS-136 phones that are capable of operating at 800 MHz and 1,900 MHz (dual-band phones) must determine the correct frequency band to scan for DCCHs prior to obtaining service. This operation must be performed quickly after power-up to achieve a fast time to service. Chapter 8 describes intelligent roaming in more detail.

3.7.1.2 DCCH scanning

Several schemes are available to help a DCCH-capable phone find service on initial power-up. Figure 3.27 provides a flowchart showing a possible DCCH scanning and locking process.

3.7.1.3 Camping criteria

There are certain signal strengths that need to be met before a DCCH is considered a suitable control channel. These are explained in Appendix C.

3.7.1.4 DCCH history list

DCCH-capable phones might maintain a history list of last-used DCCH frequencies; that is, the phone will store in memory the last channels that were used to acquire DCCH service. The phone will then scan those channels first in an attempt to find a DCCH.

3.7.1.5 Analog control channel overhead message

ACCs have an overhead message announcing the DCCH frequency; the message will directly point a DCCH-capable phone to the DCCH in that cell. If this message is not present on the ACC, the phone will

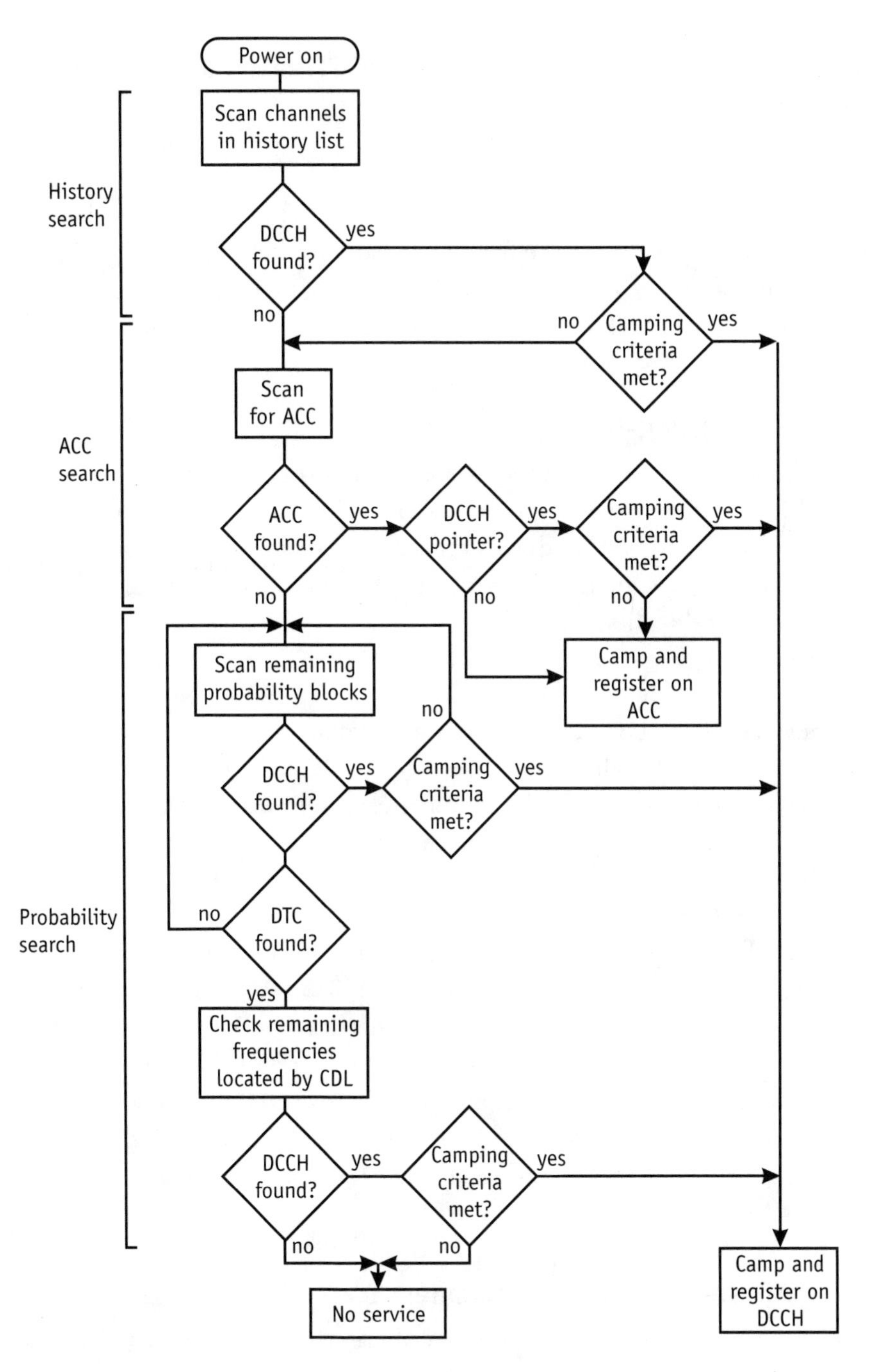

Figure 3.27 DCCH scanning and locking processes.

assume that the cell does not support DCCH operation and will then register and camp on the ACC.

3.7.1.6 Digital traffic channel locator

DTCs contain a DCCH locator in each burst. This locator will be used while a phone is scanning the frequency band looking for a DCCH. If the phone finds a DTC, it will know to decode the last 11 bits of the channel, which will point to a block of eight frequencies, one of which will be the DCCH for that cell.

3.7.1.7 Probabilistic-channel assignment

Probabilistic-channel assignment is a process that can be used to find a DCCH in a system where no ACCs are present—for example, in the 1,900-MHz frequency band. If, after searching the last-used channels and scanning the ACCs, no DCCH is located, the phone may start a scan of the frequency band to find a DCCH.

3.7.1.8 Call release message

Phones are given DCCH frequency information on analog and digital call releases. Each call-release message to DCCH-capable phones in a DCCH-equipped system will contain a channel number that will direct the phone to the serving DCCH in that cell.

3.7.1.9 Acquiring service on a DCCH

After satisfying the signal-strength requirements for camping on a DCCH, a phone must read the FBCCH prior to acquiring service. The phone has to read a full cycle of the broadcast information to gain knowledge of the system type, registration abilities, features supported, and neighbor information. If required, the phone can register and then enter a camping state, where it will be available to make calls, receive calls, register, receive teleservice messages, scan for new control channels, or be authenticated.

3.7.2 System and control channel reselection

In the AMPS cellular system, phones only reassess the best ACC for service when the signal strength of the serving control channel drops below a certain threshold, or at approximately five-minute intervals.

When this threshold is reached, analog phones generally scan the predetermined block of 21 ACCs and enter the idle state on the strongest control channel to register and to receive pages for incoming calls. Therefore, only the strongest new ACC is the viable candidate when the signal strength on the serving cell falls below a satisfactory level.

This situation might cause a problem when the system designer is attempting to make phones prefer a microcell that is at a lower transmit power than an umbrella macrocell, or when more control is required between cells in high traffic areas. In addition, microcells might not carry the traffic they were intended to carry if they cannot be acquired by phones.

Digital control-channel reselection is the function that allows a DCCH-capable phone to perform the following functions.

- Scan nonstandard control channels: Since a DCCH can be anywhere in the cellular band, the reselection process uses broadcast neighbor lists in each cell to tell a phone which neighboring DCCHs are to be found.

- Scan neighbor DCCHs in real time: Scanning is ongoing, not just when the serving signal strength degrades. This *reselection* enables a phone to build up a more accurate picture of its environment by performing more frequent evaluations of the surrounding neighbor channels.

- Make decisions on how to treat each neighbor: Based on broadcast parameters, a DCCH phone might opt to gain service from a cell that is not the strongest but is a sufficiently strong neighbor. This is the basis of a hierarchical cell structure (HCS).

In this way, reselection can be defined as the change of control channel during the camping state and can be compared to hand-off, which is the change of traffic channel during a call.

It is important to note that reselection can occur between two DCCHs or from a DCCH to an ACC. Reselection from an analog control to DCCH case is not available, since neighbor lists are not supported on the ACCs. However, a phone will use the DCCH pointer on the ACC to retune to a DCCH.

3.7.3 Hierarchical cell structures (HCS)

Since a geographical area might be covered by a mix of macrocells and microcells as well as public and private systems in a DCCH environment, an HCS has been introduced between neighboring cells in IS-136. A DCCH-capable phone will be able to reselect a particular control-channel neighbor cell over another based on the type of relationship defined between the serving cell and a neighbor cell.

The HCS designations are used by a phone to assess the most suitable control channel on which to provide service, even if the signal strength of a neighbor is not the highest being received by the phone but is of a sufficient level to provide quality service.

3.7.3.1 Preferred, regular, and nonpreferred cells

HCS enables the DCCH to identify and designate neighboring cells into three types: preferred, regular, and nonpreferred.

3.7.3.1.1 Preferred A preferred cell type has the highest preference. A hand-off (generated by the system) or reselection (generated by the phone) will be made to the preferred neighbor even if the signal strength received from the neighbor is lower than the serving cell. The main criteria here is that the preferred neighbor cell must have signal strength defined by the system designer as sufficient to provide quality service.

3.7.3.1.2 Regular A regular cell type has the second-highest preference. A hand-off or reselection will occur if the received signal strength of this neighbor is greater than the current serving cell signal strength plus a hysteresis value, and there is no eligible preferred cell.

3.7.3.1.3 Nonpreferred A nonpreferred cell type has the lowest preference. Hand-off and reselection will take place if the received signal strength of the serving cell drops below a certain threshold to provide service and if the signal strength of the neighbor is greater than the current cell plus a hysteresis. ACCs can also be specified in an HCS but the phone will, at that point, drop out of the DCCH environment. The hand-off and reselection criteria are explained mathematically in Appendix D.

It should be remembered that a cell type (either regular, preferred, or nonpreferred) is a relative attribute—a cell can be regular to one cell and preferred to others. This would be the case when two microcells are serving a WOS system and marked regular to each other. They may, however, be marked preferred to the surrounding macrocells, in which case each cell is considered regular by some neighbors but preferred (or even nonpreferred) by others.

3.7.3.2 Hierarchical cell structure example

Table 3.4 shows the neighbor-cell relationships that have been defined for the system and that will be broadcast on each cell to identify its neighbor cells. Each time the phone reselects a new DCCH, a new neighbor list is received that will tell the phone how to treat each of the new surrounding cells during reselection.

In Figure 3.28, using the broadcast neighbor information in Table 3.4, the phone camping on macrocell 1 would know that macrocell 2 was a regular neighbor and it would use the *regular neighbor cell* criteria when evaluating the received signal strength from macrocell 2. This would also mean that macrocell 2 would have to be of a greater signal strength, plus a hysteresis, to be considered a new control or traffic channel.

In comparison, the office microcell is marked as a preferred neighbor on the neighbor list broadcast from macrocell 1. If the signal strength received from the office microcell was above the predetermined *sufficient* threshold, a reselection or hand-off to that microcell would take

Table 3.4
Neighbor-Cell Information

Cell	Neighbor-Cell Relationship
Macrocell 1	Macrocell 2 as a regular neighbor Office microcell as a preferred neighbor
Macrocell 2	Macrocell 1 as a regular neighbor Office microcell as a preferred neighbor
Office microcell	Macrocell 1 as a nonpreferred neighbor Macrocell 2 as a nonpreferred neighbor

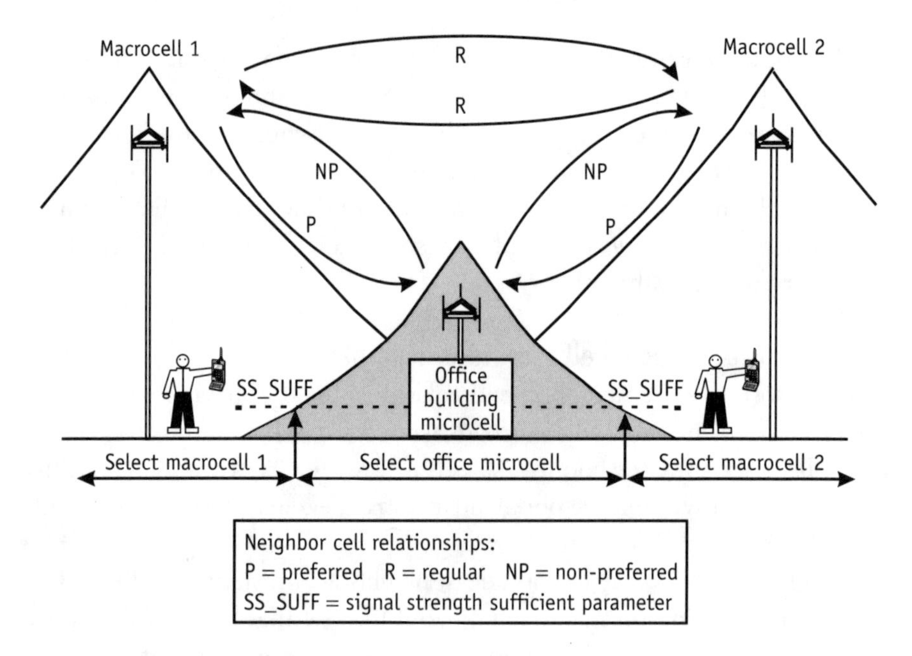

Figure 3.28 Reselection based on HCS cell designation.

place—even if the signal strengths from the macrocells were greater. This threshold, called SS_SUFF, is broadcast along with each neighbor. Appendix D describes SS_SUFF in detail.

To complete the HCS example, consider the phone as being camped on the microcell. The neighbor list would indicate both macrocells as *nonpreferred*. As the phone leaves the microcell environment—assuming the phone to be traveling left to right—the signal strength from the microcell would decrease, and the signal from macrocell 2 would increase. Using the *nonpreferred* neighbor criteria, the following two criteria must be met:

- The signal strength from the microcell would have to drop below the sufficient signal-strength threshold before a reselection or hand-off to macrocell 2 could take place.
- The signal strength from macrocell 2 would have to be of a greater signal strength, plus a hysteresis, to be considered a new control or traffic channel.

This example demonstrates that HCS is a very powerful tool for RF engineers to use with a WOS or other application where traffic needs to be managed in a microcell environment.

3.7.3.3 Neighbor cell list

Each sector that supports a DCCH can broadcast information for as many as 24 neighbor cells. This information is used by the phone when evaluating each neighbor control channel in conjunction with the neighbors' signal strength. Neighbor cells from different hyperbands may be included. That is, 800-MHz control channels can reference 1,900-MHz neighbor cells and vice versa. This allows reselection between the frequency bands and the associated seamless service.

Note that a DCCH can broadcast information regarding analog control-channel neighbors as well as DCCH neighbors.

Table 3.5 shows a summary of the neighbor-cell list information that is broadcast for each DCCH neighbor cell. A more detailed description of the parameters can be found in the IS-136 specification.

The signal strength on neighbor channels will be measured at regular intervals defined by the broadcast parameter SCAN_INTERVAL. This parameter defines the number of hyperframes between signal-strength measurements.

Table 3.5
Neighbor-Cell List Information

Parameter	Function
Channel number	RF channel number of that neighbor
Cell parameters	Network type (public, private, residential) Cell type (preferred, nonpreferred, or regular)
Cell hysteresis thresholds	IS-136 cell threshold and hysteresis parameters RSS_ACC_MIN, MS_ACC_PWR, RESEL_OFFSET, SS_SUFF, DELAY, HL_FREQ (see Appendix D)
PSID indicators	Optional information regarding private system identities
DVCC	Distinguish the channel from a cochannel

3.7.4 System registration

Registration is a function that gives a cellular system the ability to know the location and status of a cellular phone. Existing AMPS and TDMA registration schemes remain the same under IS-136, and several new forms of registration are introduced, providing for backward compatibility with existing registration schemes as well as enhanced tracking of phones' whereabouts.

For DCCH, registration messages are sent from the phone to the base station on the RACH. The registration-accept message is sent from the base station back to the phone on the ARCH.

Existing registration types maintained in IS-136 are listed as follows.

- Power-up;
- Periodic;
- Location area.

Additional types of registration supported in IS-136 are as follows.

- Power-down;
- New system registration (using SID/PSID);
- Change of control channel (from ACC/DCCH);
- Forced registration;
- User group registration;
- New hyperband;
- TMSI timeout;
- Deregistration;
- Location area using *virtual mobile location area* (VMLA).

3.7.4.1 Power-Up registration

Power-up registration is used to inform the network that a subscriber is active when the phone is powered on, allowing the network to track the activity status of the phone.

3.7.4.2 Power-down registration

Power-down registration is used to change the subscriber status from active to inactive as the phone is powered down. Phones marked inactive are not paged for call termination, thereby minimizing congestion. Power-down registration also makes it possible for the system to quickly determine when phones have powered down, enabling the system to initiate actions like call forwarding, voice mail, and recorded announcements, without having to wait for ringing time-out.

3.7.4.3 Periodic registration

Periodic registration is used to help the network keep track of the phone throughout the system. It is used on both the ACC and on the DCCH. Periodic registration can either be based on a clock sent from the system or based on an internal clock in the phone. A phone sends a registration message to the system after a timer expires to keep the phone active in the system.

3.7.4.4 New system registration (SID/PSID)

When a phone tuned to an ACC retrieves a registration ID message, it compares the current SID with its stored SID, that is, the SID corresponding to its latest-sent registration access. If the system area identities are different, a registration-access message is sent to the system.

The IS-136 specification supports the concept of private networks with a *private SID* (PSID). Phones tuned to a DCCH receive the list of private systems supported by a certain cell in the system-identity message and can register in an area served by a public/private network to which they subscribe. This registration takes place when a PSID stored in the phone matches the PSID of the public/private network broadcast on the DCCH. This registration enables a phone to keep the network informed about the system area in which it is operating. For example, the system can use this registration as a trigger to know that a phone has entered a wireless office service (WOS).

Figure 3.29 shows an example of SID and PSID change registration. The solid registration arrows indicate a change of SID registration where, like today, the phone recognizes that the SID from the new cell is different from the last cell. The dotted arrows entering the shaded area

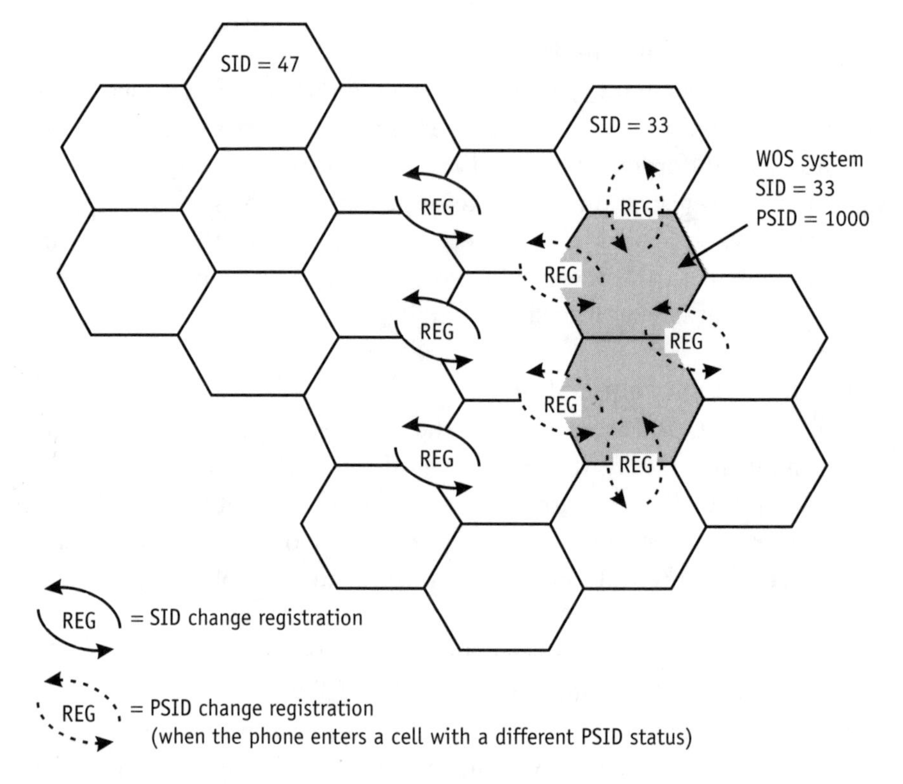

Figure 3.29 Example of SID and PSID change registration.

represent registrations when a phone reselects into a cell that is broadcasting a PSID that the phone has stored in its memory (plus other associated identifiers). This registration would be performed by phones entering their WOS area. The dotted arrows leaving the shaded WOS represent the registration that a phone performs when it leaves an area that is broadcasting a different PSID or no PSID at all.

3.7.4.5 Change of control channel type registration

An IS-136 phone can access a system either on an ACC or a DCCH. A phone that had previously accessed an ACC, and that is now acquiring service on an DCCH, is required to perform a change of control channel type registration. The phone will also reregister when the ACC is reselected after service has been acquired on a DCCH. In this way, the system knows on which control channel to page the phone.

3.7.4.6 Forced registration

It is also possible for the network to force phones to register. Forced registration allows systems to force a phone that is camping on a given DCCH to register on demand.

3.7.4.7 User group registration

User groups can be defined for point-to-multipoint paging and registration.

3.7.4.8 New hyperband

New-hyperband registration provides seamless phone operation between the 800-MHz and 1,900-MHz frequency bands for PCS services.

3.7.4.9 TMSI timeout registration

TMSI timeout registration is used when a TMSI expires.

3.7.4.10 Deregistration

Deregistration is a registration scheme through which a mobile notifies the system of its intent to leave its current network and reacquire service in a different type of network.

3.7.4.11 Virtual mobile location area

VMLA is a registration scheme that allows individual registration areas to be defined. It is based on the concept of a phone being sent a list of *registration numbers* (RNUMs) upon initial registration. Each RNUM value is associated with one or more cells in such a way that the full RNUM list defines a domain of cells.

The phone will reregister when an unrecognized RNUM is encountered on a cell. This scheme can be applied on a per-phone basis and can more accurately track and control mobile registrations. Another benefit of VMLA-based registration is that it can be used to eliminate the ping-pong registration problem by dynamically centering each new registration area around the mobile. Figure 3.30 indicates a variable (RNUM) broadcast by a system on the BCCH channel.

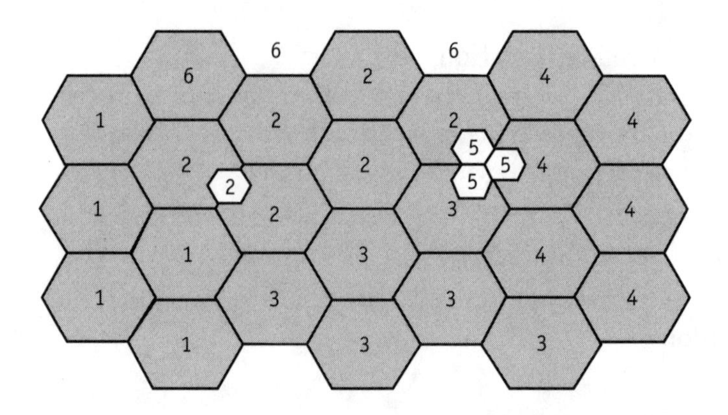

- The RNUM list is sent to phones at registration in a registration-accept message.
- Each cell broadcasts an RNUM.
- The phone will re-register if an unknown RNUM is received. This reduces problems involved with system border paging.

Figure 3.30 Registration: VMLAs.

3.7.5 Mobile-assisted channel allocation (MACA)

A new function in IS-136 similar to MAHO is MACA. MACA is a process in which signal-strength reporting takes place while the mobile phone is monitoring a DCCH (camping). While monitoring the DCCH, the phone also measures signal strengths on specified frequencies and the signal quality of the current downlink DCCH. MACA signal-strength measurements can be used for quality-based frequency assignment of a DTC (with adaptive channel allocations) and reassignment of a DCCH if the DCCH is subject to severe interference. The base station, using the BCCH, sends a neighbor list informing the phone of where to look for potential cell reselection.

The mobile phone continuously reports MACA information to the base station to assist in determining the best channel. The mobile phone performs two MACA-related functions: *long-term MACA* (LTM) and *short-term MACA* (STM). LTM is a set of data containing the word error rate, bit error rate, and received signal strength for the current DCCH. STM contains received signal strength for the current DCCH and possibly for other channels. MACA reports are also sent back to the base station during other activities such as call origination, page message responses, and registrations.

3.7.6 Call processing

To make and receive calls, wireless phones must exchange information with the system prior to service. Additionally, the phone must perform other tasks such as handoff, authentication, and teleservice transportation.

3.7.6.1 Origination

A call from a cellular phone begins with an access to the system on the control channel of the serving cell. The access includes an origination message containing the phone identification number (MIN), the serial number, and the called number. The system verifies the authenticity of the phone, the validity of the called number, and the availability of the network resources to handle the call prior to selecting an idle voice channel in the serving cell and ensures that adequate signal strength is received from the phone.

The voice channel can be analog or digital depending on the phone capability, preference, and channel availability. An IVCD message is then sent to the phone. This message, which can be repeated more than once, provides the phone with the selected voice channel to be used to complete the setup procedure.

Reorder, directed retry, and intercept messages are used to control situations that might have made the origination attempt fail.

3.7.6.2 Paging

A call to a phone begins with a page message containing the phone identification number (MIN) being broadcast over all cells in the location area where the phone is expected to be.

When this message is recognized by the appropriate phone on an ACC, the phone tunes to the strongest control channel and returns a page response message as confirmation. The page response message returns the phone identification number and, if required, its ESN.

A phone camping on the DCCH does not have to retune to the strongest control channel, but it immediately responds to the page on the current RACH since a DCCH-capable phone is always monitoring the most appropriate control channel.

If the page response is valid and meets the signal strength criteria, and if the serial-number check is active and the returned and stored serial numbers are found to be equal, then the system selects an idle analog voice or DTC in the cell identified by the page response. The voice channel is started and an initial voice-channel designation message is sent to the phone to complete the setup procedure. If no page response message is received within the paging period set by the switch, the page is considered unsuccessful.

3.7.6.3 Hand-off

IS-136 allows for hand-offs between any combination of analog voice channels and DTCs at 800 MHz, and DTCs between the 800-MHz and 1,900-MHz frequency band. The HCS algorithms may be taken into account in the hand-off process to provide coherent control and voice-channel borders within the system.

3.7.6.4 Call release

At call release, a phone is informed of the DCCH's location for its current serving cell. This message enables a phone to return to a serving DCCH directly after a call without having to rescan for a DCCH. The message is sent to DCCH-capable phones on call completion from an analog voice channel and DTC.

3.7.7 Authentication and privacy

The authentication process uses messaging on the DTCs, the DCCH, and analog voice or control channels to convey authentication information. This is used to validate a phone on the system. Authentication is a vital tool to inhibit cloning, in which a counterfeit phone takes on the identity of a normal phone for the purposes of fraudulent activity.

3.7.7.1 Message encryption

To protect sensitive subscriber information, and to enhance the authentication process, IS-136 supports encipherment of a select subset of signaling messages (those that contain the sensitive information).

3.7.7.2 Voice privacy

IS-136 provides a degree of cryptographic protection against eavesdropping in the phone-base station segment of the connection. Requests to activate/deactivate the voice privacy feature may be made during the call setup process, or when the phone is in the conversation state.

3.7.7.3 Enhanced privacy and encryption

The next revision of IS-136 will introduce an enhanced voice privacy feature, signaling message encryption and user data encryption increasing the degree of cryptographic protection.

3.7.8 Teleservice transport

IS-136 provides a mechanism for the transport of teleservice information between a teleservice server/message center in the network, and the phone. As well as standard teleservices such as CMT, OATS, OAP, and GUTS, IS-136 allows carrier-specific teleservices to support other non-standardized services.

Regardless of the teleservice being transported, IS-136 uses the R-data packets to send and receive the information. Corresponding acknowledgment packets are used to indicate the success of the delivery.

Teleservice messaging is not limited to network-originated messages. Mobile-originated teleservices provide a similar transport mechanism but from the phone or wireless device in a DCCH environment.

Future IS-136 revisions will also support a *segmentation and reassembly* (SAR) function, which will allow longer teleservice messages to be communicated by splitting the message into manageable units prior to transmission and joining them together again at the receiver.

3.7.8.1 Teleservice point-to-point delivery and acknowledgment

Figure 3.31 depicts the point-to-point delivery and acknowledgment process between the base station and the phone. The phone receives notification of a pending teleservice message (a SPACH notification) and responds with a SPACH confirmation (similar to a page and page response for voice calls). The system then begins delivery of the teleservice message that was received from the message center/teleservice server. At the end of the message delivery across the air interface, the phone sends back an accept or reject message to indicate success or failure of the message delivery.

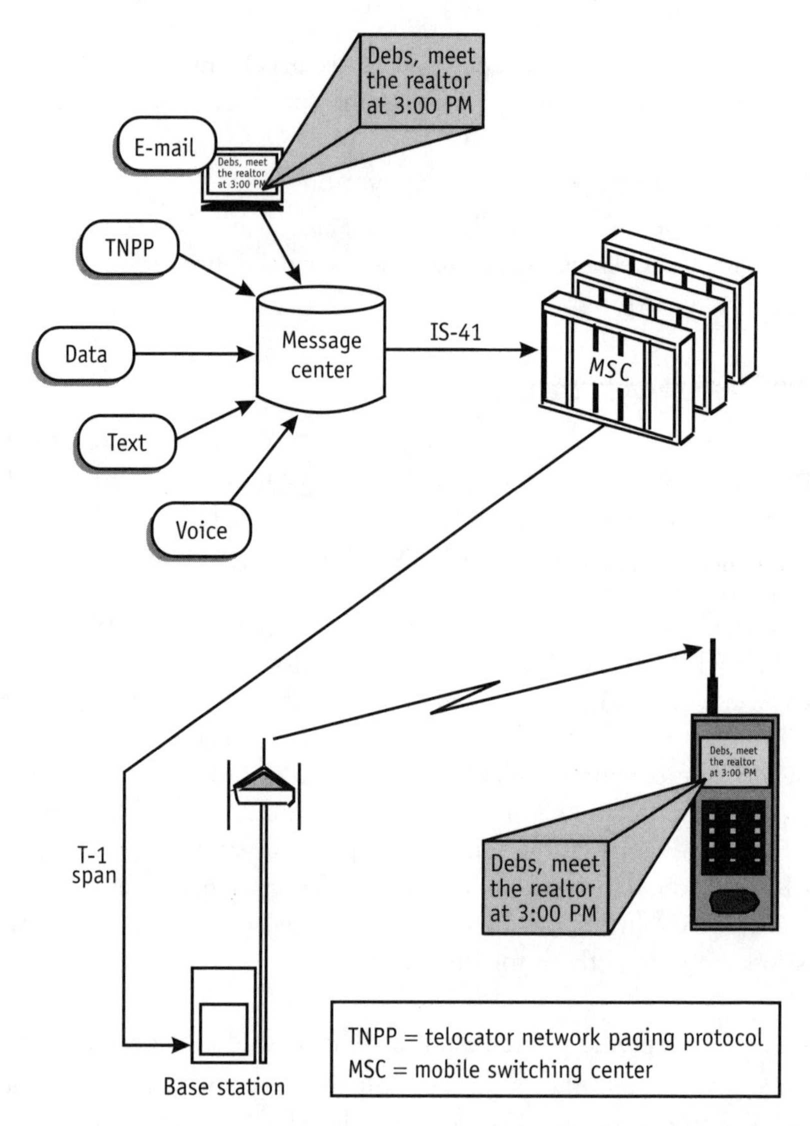

Figure 3.31 Teleservice point-to-point delivery and acknowledgment.

Each message can be acknowledged at various stages during transfer to ensure the successful delivery of over-the-air interface messages. This process includes SPACH layer-2 ARQ, layer-3 R-data accept-or-reject messages, and teleservice specific delivery and manual acknowledgments.

3.7.8.2 Teleservice delivery on the digital traffic channel

Point-to-point teleservice messages may also be delivered over the DTC using FACCH messaging. The same features and functionality of DCCH-delivered messaging are therefore included to provide a means of sending teleservice information during a voice call.

3.8 Identity structure

Private system identities, cell identities, and new *system operator codes* (SOCs) have been introduced to support the new features and capabilities on the DCCH. The IS-136 system allows for several different types of phone identities. The traditional use of 10-digit directory numbers as a means of uniquely referencing mobile phones has been augmented to provide increased paging capacity, thereby allowing for full international roaming while remaining backward-compatible with today's phone identity strategy.

3.8.1 Mobile phone identities

There are four key methods for identifying a phone.

3.8.1.1 Temporary mobile station identity

A TMSI is a 20- or 24-bit number representing a temporary mobile identity that is assigned to a phone by the system at initial registration. This shortened identifier provides enhanced paging capacity on the air interface, since a single page of messages can reference more 20-bit TMSIs than 34-bit MINs and, thus, page more phones at a time.

Table 3.6

SID Structures: Network Types

Network Type	Value
Public	1XX
Private	X1X
Residential	XX1

3.8.1.2 International mobile station identity (IMSI)

An IMSI is an internationally unique number developed to facilitate seamless roaming in future global mobile networks. Its length is 50 bits.

3.8.1.3 Mobile identification number (MIN)

Today's MIN is a 34-bit number. It continues to serve as the initial MSID, enabling the current networking and authentication mechanisms to coexist with the DCCH.

3.8.1.4 Electronic serial number (ESN)

The ESN is an 11-digit serial number that is unique to a mobile phone.

3.8.2 System identity structures

New methods of identifying IS-136 systems have been introduced to support enhanced features such as private systems.

3.8.2.1 System identity (SID)

The SID has the same meaning in IS-136 as in an AMPS system and represents an international identification and a system number identifying the service area and frequency band (A or B 800 MHz, A through F 1,900 MHz).

3.8.2.2 Network types

A new SID structure has been introduced to allow cellular phones to distinguish between public, private, semiprivate, and residential (personal) base stations (see Table 3.6). These new network types allow phones to behave differently according to the type of system providing service to the user. For example, phones only requiring plain old cellular service need not reselect or camp on cells marked private, thereby improving their time to service. Also, phones requiring service on a residential system might perform different scanning routines in order to find their home system. The network type is represented by a 3-bit field on the DCCH structure and by neighbor-cell messages on the broadcast DCCH.

A DCCH might take on the identity of several network types at the same time since X represents the do not care setting. This is useful when a public macrocell is also used to provide a private WOS to a customer. In this case, the cell would be classed as semiprivate. In the DCCH environment, cells can also be defined within several categories of network types, including the following:

- *Public:* Cells that provide the same basic service to all customers;
- *Semiprivate:* Cells that provide the same basic service to all customers and provide special services to a predefined group of private customers. This type would be used in the case of a cell providing service to a WOS system as well as to public users.
- *Private:* Cells that provide special services to a predefined group of private or WOS customers only and that do not support public use of that cell.
- *Semiresidential:* Cells that provide the same basic service to all customers and special services to a predefined group of residential customers. This type would be used in a neighborhood where the public macrocell was also providing residential cellular service.
- *Residential:* Cells that provide special services to a predefined group of residential customers only and that do not support public use of the cell. The *personal base station* (PBS) would be classed as a residential system.
- *Autonomous:* Cells that broadcast a DCCH in the same geographical area to other DCCH systems but that are not listed as a neighbor on the neighbor list of the public system. Examples of autonomous systems would be the personal base station (PBS) or private microcell systems that are not frequency-coordinated with the public system. These cells require the phones to perform special frequency-scanning algorithms in order to find them.

3.8.2.3 System operator code (SOC) and base station manufacturer code (BSMC)

The SOC and the BSMC are new system identities that enable a phone to recognize base stations either belonging to a certain cellular operator (SOC) or supplied by a specific manufacturer (BSMC). In addition to system recognition, these identities enable a phone to activate proprietary signaling protocols to provide advanced services that might not be available from other carriers or equipment manufacturers. IS-136 defines the SOC and BSMC assignments.

3.8.2.4 Mobile country code (MCC)

A *mobile country code* (MCC) is included in system broadcast information to identify the country in which the system is operating. This supports international applications of IS-136 and international roaming.

3.8.2.5 Private system identity (PSID)

A PSID is assigned to a specific private system by the operator to identify that system to phones in the coverage area of that system. PSIDs can be assigned on a sector-by-sector basis that allows very small service areas to be defined. Alternatively, many cells, as well as systems, could broadcast by the same PSID to create a geographically large virtual private system. Phones that recognize PSIDs will notify the system and display a specific system name on the screen to inform the user that they have entered the private system.

A single DCCH can broadcast up to sixteen PSIDs, allowing the support of up to sixteen different virtual private systems on one DCCH. This feature would be useful in a technology park or campus where it would not be economical to support a DCCH for each small customer requiring WOS features. Figure 3.32 shows a typical private system configuration.

It is important to note that a default system banner or alpha tag of the public system is displayed when the phone is not near the location of its private system. This is shown in Figure 3.32. The phone is looking for a PSID of 9927 (Mountain High Ski Resort) but that PSID is not broadcast for this cell. In this case, the phone would display the default alphanumeric SID of the public system, as shown in Figure 3.32.

3.8.2.5.1 PSID Ranges There are over 65,000 different PSIDs split into four ranges, with each range having unique properties described in the following sections.

Range 1. Range 1 is reserved for SID-specific PSIDs, that is, PSIDs that will be meaningful in only one service area. An example is a small business WOS system that is contained in one or more sites under a single SID

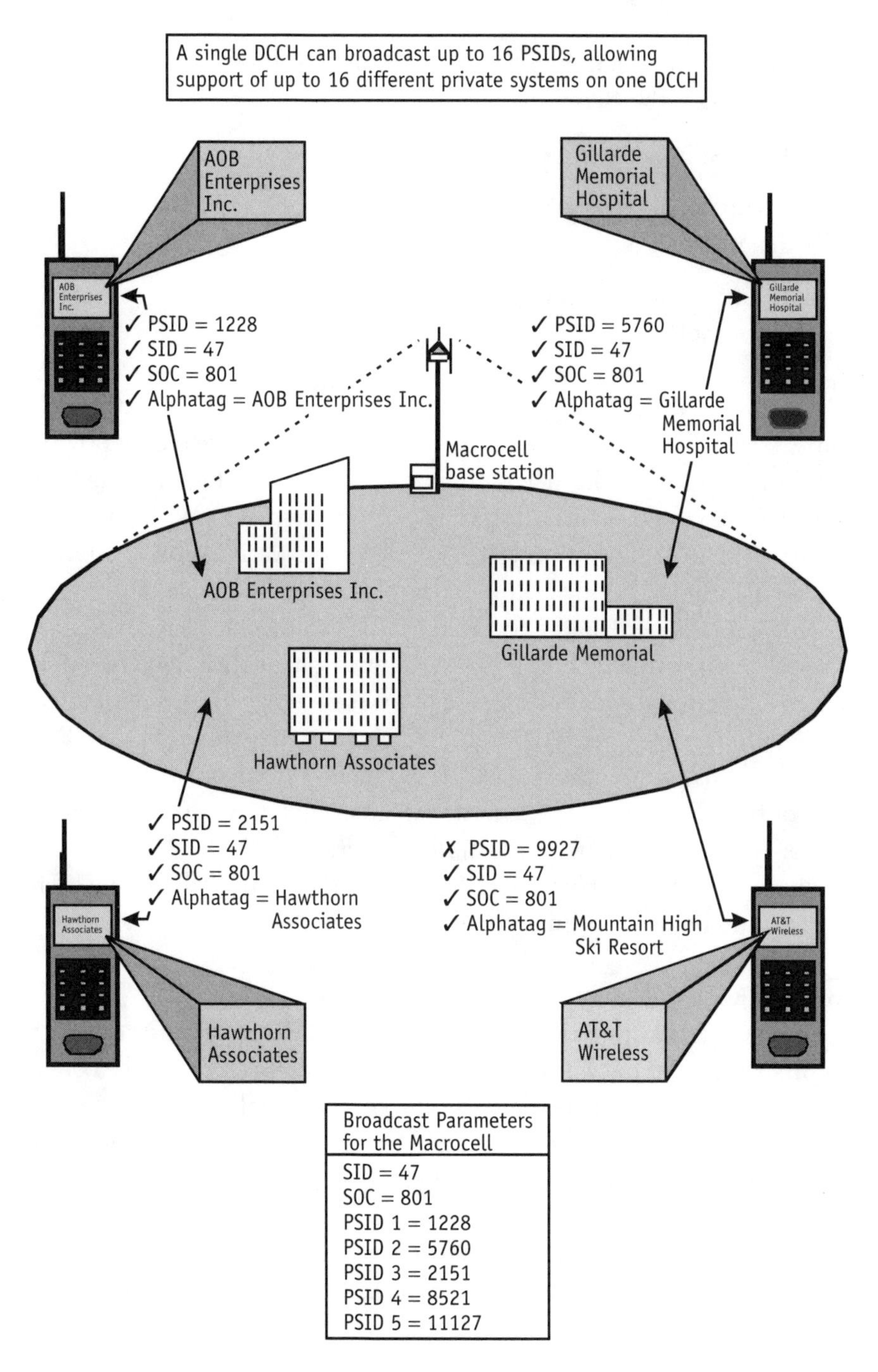

Figure 3.32 Private system configuration.

service area, although it does not cover multiple cities. Reasons for using Range 1 include the following.

- The PSIDs are wholly dependent on the SID.
- No intermarket negotiation needs to take place to allocate PSIDs in this range.
- The PSIDs in this range can be safely reused in any market that has a different SID.

Range 2. Range 2 has significance for PSIDs related to national accounts served by a single cellular carrier. These PSIDs are independent of the SID and have meaning across all markets broadcasting the same SOC. These PSIDs would be used for large national accounts with customers who roam all over the country. This designation would allow the phones to be programmed in one home market for use in multiple cities.

Range 3. Range 3 takes the PSID designation one step further and provides a range in which PSIDs are relevant on a nationwide basis. For example, a national account receiving service from multiple cellular carriers would use the same PSID for all cellular service provided. The PSID defined in this range would have to be standardized on an intercarrier basis but would allow large national customers to receive seamless private-system service from multiple carriers.

Range 4. Range 4 provides a PSID range that is globally unique. This range would be used for international roamers.

3.8.2.6 Residential system identity (RSID)

In a manner similar to PSIDs, *residential SIDs* (RSIDs) identify residential systems within the public cellular coverage. RSIDs can be used to create residential-service areas or neighborhood residential systems by broadcasting an identifier that is recognized by phones as being at home and, therefore, receiving special services (such as billing). A primary use of RSIDs is in the PBS, which allows a cellular phone to be used like a cordless phone in conjunction with a residential cellular base station. RSIDs can also be broadcast on the macrocellular system to create virtual residential networks.

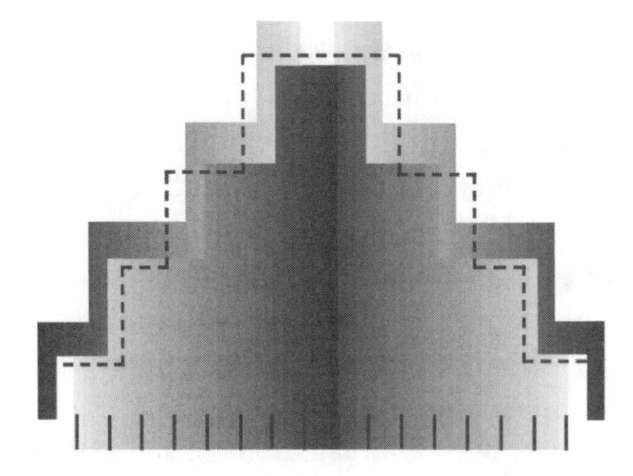

4

TDMA Mobile Telephones

Mobile telephones are the link between the customer and the wireless network. Mobile telephones must provide a method for the customer to control the phone and display the operating status. Phones must also sample and process audio signals, transmit and receive radio signals via an RF section, and obtain energy—typically from a battery—to operate the complex electronics. The basic parts of a mobile phone include the following.

- *Man-machine interface* (MMI)—display and keypad;
- *Radio frequency* (RF) section;
- Signal processing (audio and logic);
- Power supply/battery.

Mobile telephones may be mobile radios mounted in motor vehicles, transportable radios (mobile radios configured with batteries for out-of-the-car use), or self-contained portable units. The official name of a mobile telephone from the IS-136 industry standard specification is *mobile station.*

In addition to the key assemblies contained in a mobile phone, accessories must be available that include battery chargers, hands-free assemblies, and data adapters. These accessories must work together with the mobile phone as a system. For example, when a portable mobile phone is connected to a hands-free accessory, the mobile phone must sense that the accessory is connected, disable its microphone and speaker, and route the hands-free accessory microphone and speaker to the signal processing section.

The IS-136 system allows mobile phones to operate in multiple modes (analog mode and digital mode) and dual frequencies (800 MHz and 1,900 MHz). Figure 4.1 illustrates the functional sections of an IS-136 dual-mode and dual-band mobile phone.

4.1 Man-machine interface

Customers control and receive status information from their mobile phones via the MMI. This interface consists of an audio input (microphone) and output (speaker), a display device, a keypad, and an accessory connector to allow optional devices to be connected to the mobile phone. The mobile phone's software coordinates all these MMI assemblies.

4.1.1 Audio interface

The audio interface assembly consists of a speaker and microphone that allow customers to talk and listen on their mobile phone. While the audio assemblies are located in a handset, they can be temporarily disabled and replaced by a hands-free accessory.

The microphone in small portable telephones is very sensitive. It allows the mobile phone to detect normal conversation even when the microphone is not placed directly in front of the speaker's mouth. This is especially important for the microportable phones that are shorter than the distance between a customer's ear and mouth.

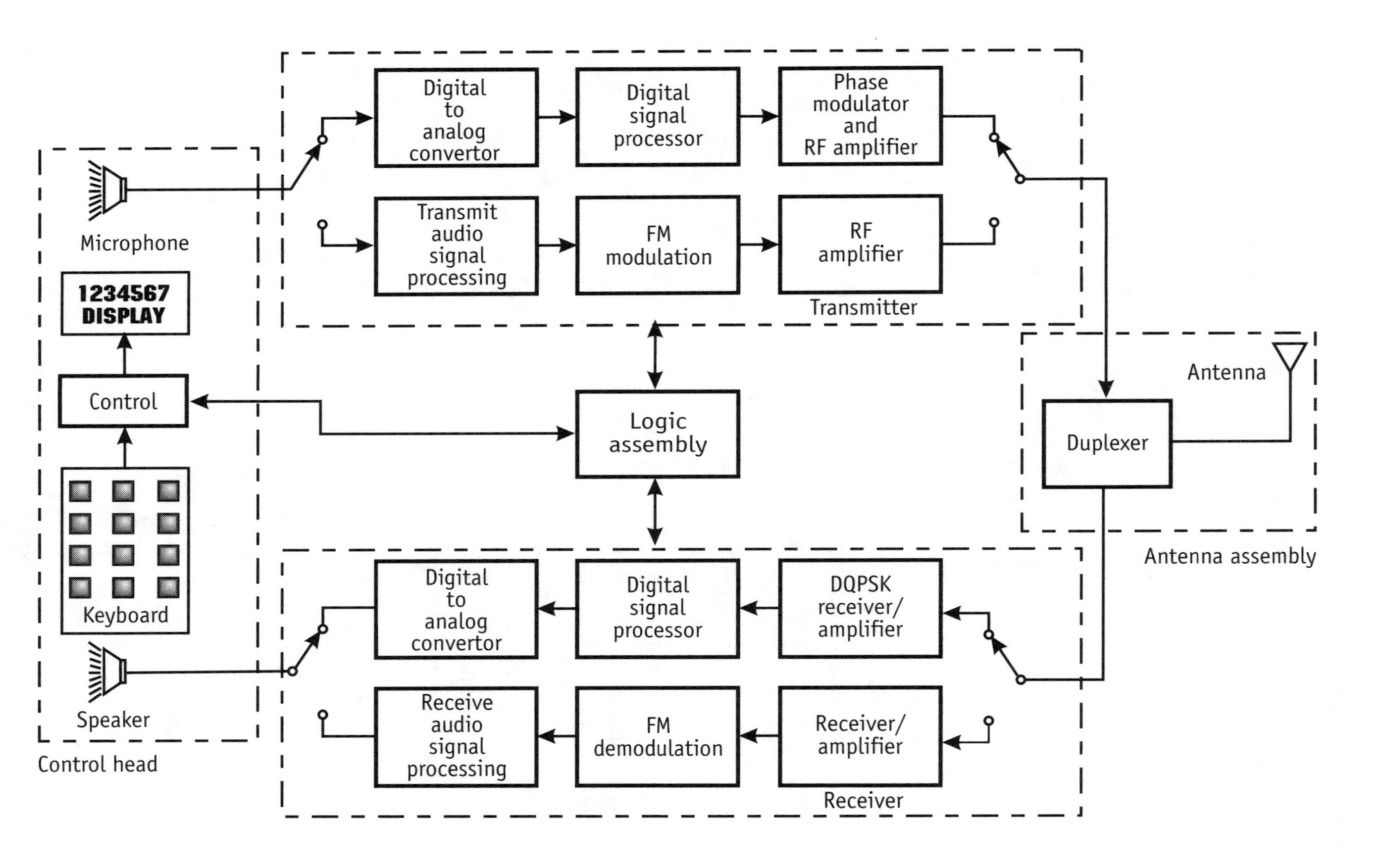

Figure 4.1 TDMA mobile telephone block diagram.

4.1.2 Display

Mobile phones typically have a display assembly that allows the customer to see dialed digits, status information, stored information, messages, and call status information such as radio signal strength. For IS-136 mobile phones, dialed digits are displayed and can be changed before the call is initiated. This is known as *preorigination dialing*.

Status indicators provide the customer with key information about their phones' operation. These status indicators typically include a "ready" indication, RF signal strength, battery level, call indication, and others. Some of these indicators are icons (symbols) or text messages. Because mobile phones must find available service prior to requested service, the display may indicate "wait" while the phone is searching for an available system. To allow customers to determine whether they are in an area with radio signal strength that is sufficient to initiate or maintain a call, an RF signal level indicator is typically provided. For portable phones, a battery level indicator may be provided to display the remaining capacity of the battery that is available to initiate or finish a call.

Most mobile phones have the capability to store and manipulate small amounts of information in an electronic phonebook. In addition to storing phone numbers, some models allow the storage of a name tag along with the number. Because many mobile phones can display eight to 12 characters across, name tags are typically limited to only a few letters.

The IS-136 system is capable of sending text messages to mobile phones. Mobile phones have several creative approaches to displaying these alphanumeric messages. Some phones show messages in *pages* one screen at a time. Other phones use a technique known as *marqueeing,* in which a message is scrolled across the screen. This allows the mobile phone that can display only a few characters per line to display lengthy messages to the customer. Alternatively, accessory display, such as the Reflection Technologies display shown in Figure 4.2, shows an accessory display that can be connected to a mobile phone to allow entire pages of text to be viewed by the customer.

4.1.3 Keypad

The keypad allows the customer to dial phone numbers, answer incoming calls, enter name tags into the phone's memory, or, in some cases, use

Figure 4.2 Accessory display. (*Source:* Reflection Technologies.)

the phone as a remote control device via the cellular system. While a keypad is typically used in mobile phones, the keypad may sometimes be replaced by an automatic dialer (auto-security) or by a voice recognition unit.

The layout and design of keypads vary from manufacturer to manufacturer. A typical keypad will contain keys for the numbers 0–9, the * and # keys (used to activate many subscriber services in the network), volume keys, and a few keys to control the user functions. In addition, keypads will contain a *send* and *end* button, which starts and terminates calls. Special function keys may also be included for speed dialing and menu features. Customers can typically access special phone options (such as the type of ringing sound or volume) via the keypad.

4.1.4 Accessory interface

There are several mobile phone accessories that can typically be attached via an accessory connector (plug). The accessory connector usually provides control lines (for dialing and display information), audio lines (input and output), antenna connection, and power lines (input and output from the battery). Accessory connections are generally proprietary, and no industry standard accessory interface connection has been agreed on for mobile phones. Each manufacturer, and often each model, will have a unique accessory interface. The types of accessories include passive devices such as external antennas, active devices such as a hands-free speaker cradle, computer-controlled devices, and various power supply options. Figure 4.3 shows a typical accessory connection.

4.2 Radio frequency section

The mobile phone's RF section consists of transmitter, receiver, and antenna assemblies. The transmitter converts low-level audio signals to modulated shifts in the RF carrier frequency. The receiver amplifies and

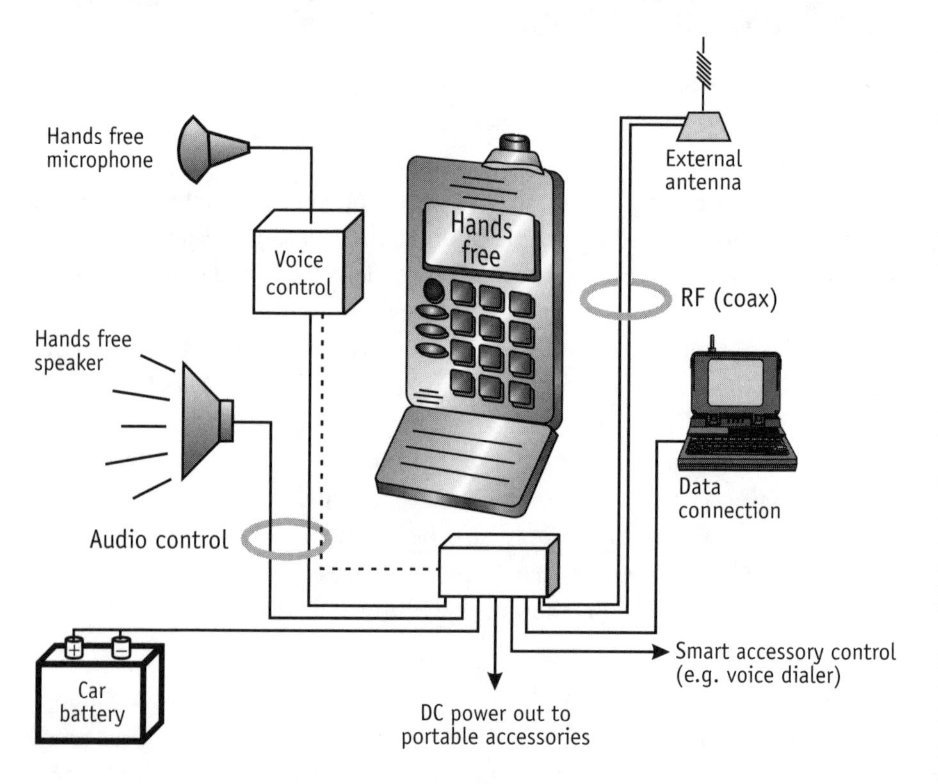

Figure 4.3 Typical accessory connection.

demodulates low-level RF signals into their original audio form. The antenna section converts RF energy to and from electromagnetic signals. There are several new requirements for the digital RF section used in IS-136 mobile phones.

4.2.1 Transmitter

The transmitter section contains a modulator, a frequency synthesizer, and an RF amplifier. The modulator converts audio signals to low-level RF modulated radio signals on the assigned channel. A frequency synthesizer creates the specific RF frequency the cellular phone will use to transmit the RF signal. The RF amplifier boosts the signal to a level necessary to be received by the base station.

The transmitter is capable of adjusting its transmitted power up and down in 4-dB or 6-dB steps depending on commands it receives from its serving base station. This allows the mobile phone to transmit only the necessary power level to be received at the serving base station, thereby reducing interference to nearby base stations that may be operating on the same frequency.

The type of RF amplification for IS-136 digital transmission is different than AMPS RF amplification. Since AMPS mobile phones use only FM and FSK modulation, they can use nonlinear (Class C) amplifiers. The IS-136 digital radio channel uses phase modulation that requires the use of linear (class A or AB) amplifiers. Fortunately, the same linear amplifier can be used for AMPS and IS-136 TDMA digital modulation.

The frequency accuracy for IS-136 mobile phone transmitters requires more precise frequency control than the control used in AMPS phones. To maintain accurate frequency control, IS-136 frequency synthesizers (generators) are locked to the incoming radio signal of the base station. If the mobile phone has the capability for dual bands of frequencies, additional filters must be used for the transmitter section to allow for both 800-MHz and 1,900-MHz channels.

4.2.2 Receiver

The IS-136 mobile phone's receiver section contains a bandpass filter, a low-level RF amplifier, an RF mixer, and a demodulator. RF signals from the antenna are first filtered to eliminate radio signals that are not in the cellular band (such as television signals). The remaining signals are sent to a low-level RF amplifier and are routed to an RF mixer assembly. The mixer converts the frequency to a lower frequency that is either directly provided to a demodulator or sent to another mixer to down convert the frequency even lower (dual stage converter). The demodulator converts the proportional frequency or phase changes into low-level analog or digital signals. If the mobile phone has the capability for dual bands of frequencies, additional filters and a high frequency mixer must be used to allow for both 800-MHz and 1,900-MHz channels.

4.2.3 Antenna

An antenna section converts electrical energy into electromagnetic energy. Although antennas are passive devices (meaning that no signal amplification is possible), they can provide a signal gain by focusing the transmission in a desired direction.

The antenna may be an integral part of the mobile phone (in a handheld portable phone), or it may be externally mounted (on the top of a car). Antennas can have a gain where energy is focused into a beamwidth area. This focused energy allows the phone to communicate over greater distances, but as the angle of the antenna changes, the direction of the beam also changes, reducing performance. For example, car-mounted antennas that have been tilted to match the style lines of the automobile often result in extremely poor performance.

If a mobile phone is connected to an external antenna, the RF cabling that connects the transmitter to the antenna adds losses that reduce the performance of the antenna assembly. This loss ranges from approximately .01 to .1 dB per foot of cable, depending on the type of cable and the RF frequency.

Typical antennas are a quarter-wavelength long. If the mobile phone has the capability for dual bands of frequencies, the antenna must be designed to operate at both 800 MHz and 1,900 MHz. Figure 4.4 shows a dual-band antenna.

4.3 Signal processing

IS-136 TDMA phones require approximately 100 times more signal processing capability than AMPS cellular phones (which typically require

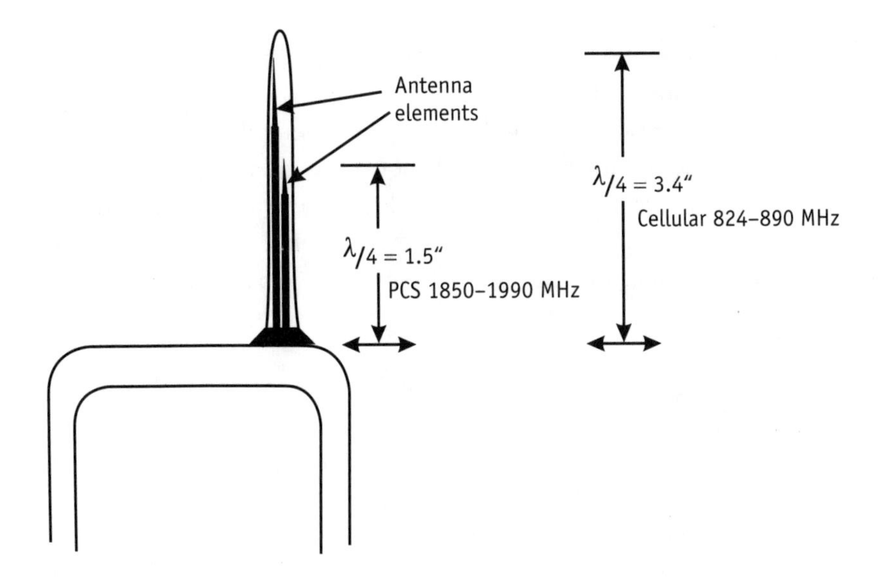

Figure 4.4 Dual-band antenna.

40 to 60 *million instructions per second* [MIP]). Most of the signals processed by digital mobile phones are in digital form. To process the digital signals, these mobile phones typically use high speed DSPs or ASICs.

Because IS-136 RF signals are converted to digital form and are demodulated using software programs, the same process can be used to modulate and demodulate analog (AMPS) or digital (DTC) signals by using different software programs. This allows a single design to be used for the IS-136 phone for both analog and digital signal processing.

The power consumption of high-speed DSPs is roughly proportional to their operating voltage and processing speed. The first DSPs consumed well over 20 to 30 mW per MIP, which presented a challenge for extended battery life in portable mobile phones The newest commercial DSP technologies consume less than 5 mW per MIP.

4.3.1 Speech compression

A significant portion of the digital signal processing used in digital mobile involves digital speech compression and expansion. This process, which is called *speech coding,* involves the conversion of analog (audio) signals into digital signals (analog-to-digital conversion). The digitized voice signal is then processed by a speech coding program to create a characteristic representation of the original voice signal (key parameters). After some error protection is added to the compressed digital signal, this information is sent via the radio channel to be transmitted. When it is received, it is recreated to the original analog signal by decoding the information using a speech decoding program. The IS-136 system can use various types of speech coding to maintain system backward compatibility while offering enhanced voice quality to customers.

4.3.2 Channel coding

The process of adding error protection and detection bits and multiplexing control signals with the transmitted information is called channel coding. Error protection and detection bits—which may be the same bits—are used to detect and correct errors that occur on the radio channel during transmission. The output of the speech coder is encoded with additional error protection and detection bits according to the channel coding rules for its particular specification. This extra information allows the receiver to determine whether distortion from the radio transmission has caused errors in the received signal. Control signals such as power control, timing advances, and frequency hand-off must also be merged into the digital information to be transmitted. The control information may have a more reliable type of error protection and detection process that is different from the speech data. This is because control messages are more important to the operation of the mobile phone than voice signals. The tradeoff for added error protection and detection bits is the reduced amount of data that is available for voice signals or control messages. The ability to detect and correct errors is a big advantage of digital coding formats over analog formats, but it does come at the cost of the additional data required.

4.3.3 Audio processing

In addition to digital signal processing for speech coding and channel coding, a digital mobile phone can do other audio processing to enhance its overall quality. Audio processing may include detecting speech among background noise, noise cancellation, or echo cancellation.

Echo is a particular problem for audio signals in digital systems. Echo can be introduced by the delay involved in the speech compression algorithm or through normal speaker phone operation. The echo signals can be removed by sampling the audio signal in brief time periods and looking for previous audio signal patterns. If the echo cancellor finds a matched signal, it is subtracted, thus removing the echo. This process may sound simple, but it is actually complicated. There may be several sources and levels of echo, and they may change over the duration of the call.

4.3.4 Logic control section

The logic control sections usually contain a simple microprocessor or microprocessor section stored in a portion of an ASIC. The logic section coordinates the overall operation of the transmitter and receiver sections by extracting, processing, and inserting control messages. The logic control section operates from a program that is stored in the mobile phone's memory.

Various types of memory storage are used in an IS-136 TDMA mobile phone. Part of the memory holds the operating software for the logic control section. Some phones use flash (eraseable) memory to allow the upgrading of this operating software to allow software correction or the addition of new features. Typically, TDMA phones contain 1 megabyte or more of memory. While operating software is typically loaded into the phone at the factory, some mobile phones can have their memory updated in the field. This feature is discussed later in the software download section (Section 4.5.3).

Read-only memory (ROM) is used to hold information that should not be changed in the phone (such as the startup processing procedures). *Random access memory* (RAM) is used to hold temporary information (such as channel number and system identifier). Flash memory is typically used to hold the operating program and user information (such as names and stored phone numbers).

4.3.5 Subscriber identity

Each mobile phone must contain identification information to provide its unique identity to wireless systems during system access attempts. IS-136 TDMA mobile phones have several unique codes.

The most basic form of identification is stored in the NAM. The NAM contains information specific to a cellular phone, such as its MIN and home SID information. In addition to the NAM information (which can be changed), there is a unique ESN that is assigned to each mobile phone. The information contained in the NAM and the ESN is used to identify the phone to a cell site and MSC.

In early AMPS mobile phones, the NAM information was stored in a *programmable ROM* (PROM) chip that required a programmer called a *NAM burner*. It was inconvenient to remove the NAM chip, which was installed by the programmer. In the late 1980s, mobile phone manufacturers designed phones to store NAM information directly in the handset without a PROM change and to allow dealers to program the NAM information directly into the mobile phone via the keypad. IS-136 provides a mechanism for the NAM contents to be downloaded directly to the phone over the air so that even the entry of the NAM data via the keypad is now unnecessary.

In addition to basic NAM information, IS-136 mobile phones store other unique identification information, including alphanumeric system identifiers and data for authentication.

4.4 Power supply

There are a variety of power supply options for IS-136 mobile phones, including batteries, converters, and chargers. There are typically several sizes (capacities) of batteries for each model of mobile phone manufactured. The capacity of the battery varies depending on its size and battery technology. Most mobile phone models also have an available battery eliminator that is used to directly connect a mobile phone to a car's cigarette lighter. When connected to the cigarette lighter, battery eliminators can also charge the battery of the phone.

There are several types of batteries used in mobile phones: alkaline, *nickel cadmium* (NiCd), *nickel metal hydride* (NiMH), and *lithium ion* (Li-Ion). In addition, Zinc Air, a new type of battery technology with increased storage capacity, is now being explored.

With the introduction of portable cellular phones in the mid 1980s, battery technology became one of the key technologies for users of cellular phones. In the mid 1990s, over 80% of all cellular phones sold were portable or transportable models rather than fixed installation car phones. Battery technology is a key factor in determining portable phones' size, talk time, and standby time.

Batteries are categorized as primary or secondary. Primary batteries must be disposed of once they have been discharged, while secondary batteries can be discharged and recharged for several cycles. Primary cells (disposable batteries) which include carbon, alkaline, and lithium have limited use in cellular mobile phones.

4.4.1 Batteries

Alkaline batteries are disposable batteries that are rarely used for mobile phones. Disposable batteries are readily available, have a very long shelf life, and do not require a charging system. However, they must be replaced after several hours of use and are more expensive than the cost of recharging a battery. Accordingly, alkaline batteries are generally not well-suited for use in cellular phones because the high current demands of the phones in transmit mode limit the useful life of the battery.

NiCd batteries are rechargeable batteries that are constructed of two metal plates made of nickel and cadmium placed in a chemical solution. An NiCd cell, which can typically be cycled (charged and discharged) 500 to 1,000 times, is capable of providing high power (current) demands required by the radio transmitter sections of portable mobile phones. While NiCd cells are available in many standard cell sizes such as AAA and AA, the battery packs used in cellular phones are typically uniquely designed for particular models of mobile phones. Some NiCd batteries develop a memory of their charging and discharging cycles, and their useful life can be shortened significantly if they are not correctly discharged. Under a phenomenon known as the *memory effect,* the battery remembers a certain charge level and will not provide more energy even if completely charged. Fortunately, newer NiCd batteries use improved designs that reduce the memory effect.

NiMH batteries are rechargeable batteries that use a hydrogen adsorbing metal electrode instead of a cadmium plate. NiMH batteries can provide up to 30% more capacity than a similarly sized NiCd battery.

However, for the same energy and weight performance, NiMH batteries cost about twice as much as NiCd batteries.

Li-Ion batteries, which are either disposable or rechargeable batteries, represent the newest technology being used in mobile phones. These batteries provide increased capacity versus weight and size. A typical Li-Ion cell provides 3.6V versus 1.2V for NiCd and NiMH cells. This means that one-third the number of cells are needed to provide the same voltage. Figure 4.5 shows the relative capacity of different battery types.

4.4.2 Battery chargers

There are two types of battery chargers: trickle and rapid charge. A trickle charger will slowly charge up a battery by allowing only a small amount of current to be sent to the mobile phone. The battery charger may also be used to keep a charged battery at full capacity if the mobile phone is regularly connected to an external power source (such as a car's cigarette lighter socket). Rapid chargers allow a large amount of current to be sent to the battery to fully charge it as soon as possible. The limitation on the rate of charging is often the amount of heat generated; the larger the amount of current sent to the battery, the larger the amount of heat.

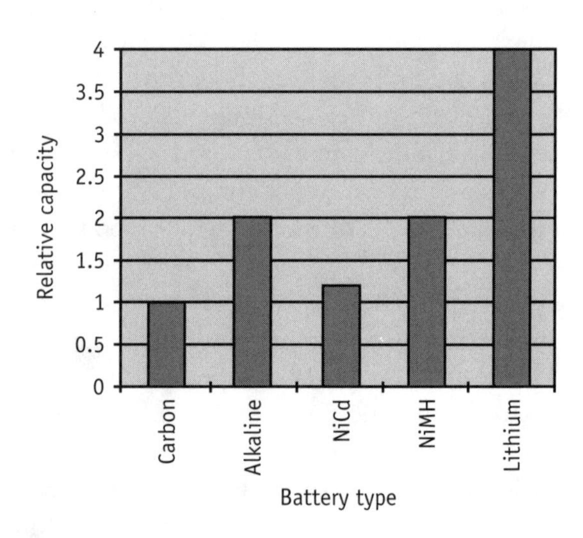

Figure 4.5 Battery storage capacity.

The charging process is controlled by either the phone itself or by circuits in the charging device. For some batteries, rapid charging reduces the number of charge and discharge cycles. A charger will charge until a voltage transient (or knee voltage) occurs and checks the battery's temperature. The full charge is indicated by a few different conditions: Either the temperature of the battery reaches a level where the charging must be turned off, the voltage level reaches its peak value for that battery type, or the voltage level stops increasing. Most chargers will then enter a trickle charge mode to keep the battery fully charged. In a process called *battery reconditioning,* some chargers for NiCd batteries discharge the battery before charging to reduce the memory effect.

4.4.3 IS-136 standby and talk time

IS-136 technology provides unique tools for decreasing phone power consumption and increasing the standby time of batteries. The operation of the DCCH is such that phones do not have to receive the control channel information continuously as they do in an AMPS system. Since the phone and system are synchronized, the phone can power off some of its circuitry while waiting between information bursts. This means that, typically, only one out of 192 bursts of information has to be read on the control channel—significantly increasing the battery life. Figure 4.6 shows the relative current consumption of an AMPS and IS-136 mobile phone.

4.5 Accessories

Accessories are optional devices that may be connected to a mobile phone to increase their functionality. Accessory devices include the hands-free speakerphone, smart accessories (modems), voice activation, battery eliminators, antennas, and many others.

4.5.1 Hands-free speakerphone

Hands-free car kits typically include a microphone and speaker to allow the subscriber to talk into the phone without using the handset. The speaker is usually located in the cradle assembly, while the microphone is

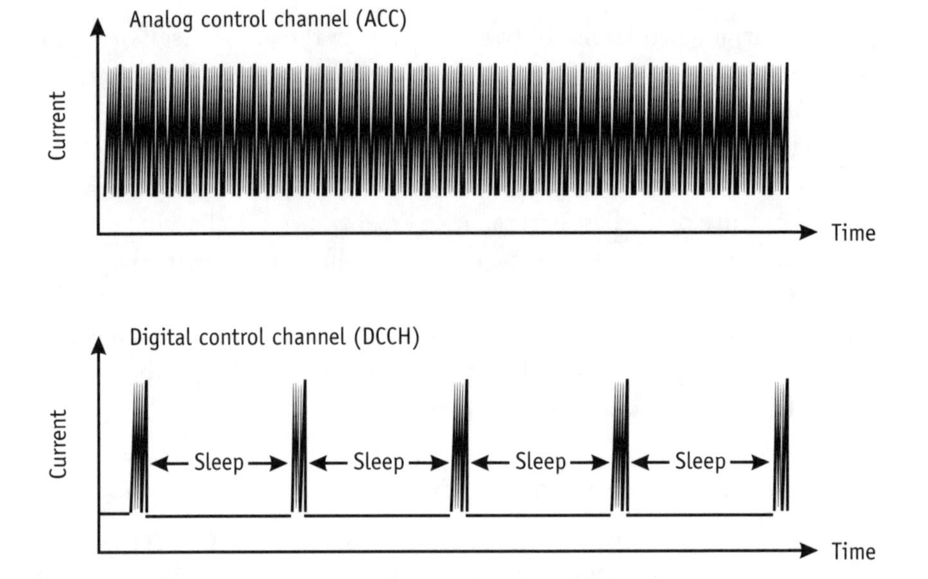

Figure 4.6 AMPS and IS-136 current consumption.

typically installed in a remote microphone, usually located near the visor. Some hands-free systems have advanced echo-canceling technology to minimize or eliminate the effects of delayed echo that can be introduced in digital systems. Figure 4.7 depicts a hands-free speakerphone.

4.5.2 Smart accessories

AMPS mobile phones are capable of connecting to various types of devices, including computer modems. When these smart accessory devices are used, they typically require an audio connection for the modem data transfer. IS-136 also includes an all-digital data service that does not require audio channels and connects directly to a computing device without a modem. Chapter 7 provides additional information on data transmission.

Another optional feature is voice activation, which allows phone users to dial and control calls with voice commands. It is recommended that mobile phone users not dial calls while driving because of safety concerns, but a mobile phone user can dial (initiate) a call via voice activation without significant distraction.

Figure 4.7 Hands-free speakerphone. (*Source:* Cellport.)

There are two types of speech recognition—speaker-dependent and speaker-independent. Speaker-dependent voice recognition requires users to store voice commands to be associated with particular commands. These recorded commands are used to match words spoken during operation. Speaker-independent voice recognition allows multiple users to control the phone without the recording of a particular voice. To prevent accidental operation of the mobile phone by words in normal conversation, key words such as "phone start" are used to indicate that a voice command will follow.[1]

4.5.3 Software download transfer equipment

Some mobile phones can have their operating system memories reprogrammed in the field to add feature enhancements or to correct software

1. U.S. Patent 4,827,520, "*Voice-Actuated Control System for Use in a Vehicle*," Mark Zeinstra, 1989

errors. The new operating software is typically downloaded using a service accessory that contains an adapter box connected to a portable computer and a software disk. (See Figure 4.8.) The new software is transferred from the computer through the adapter box to the mobile phone. Optionally, an adapter box can contain a memory chip with the new software, thereby eliminating the need for the portable computer. Phone users can make changes easily in the field without opening up a mobile phone.

4.5.4 Antennas

Antennas on portable mobile phones are typically integrated into the case of the mobile phone. Some phones include a coaxial antenna connection that allows the external antennas to be mounted on a car.

Several factors affect the performance of an antenna. Antennas convert radio signal energy to and from electromagnetic energy for transmission between the mobile phone and base station. Because there is only a fixed amount of energy available for conversion, antennas can only improve their performance by focusing energy in a particular direction that reduces energy transmitted and received in other directions. The amount of gain is specified relative to a unity (omnidirectional) gain antenna. Car-mounted antennas typically use 3- or 5-dB gain antennas.

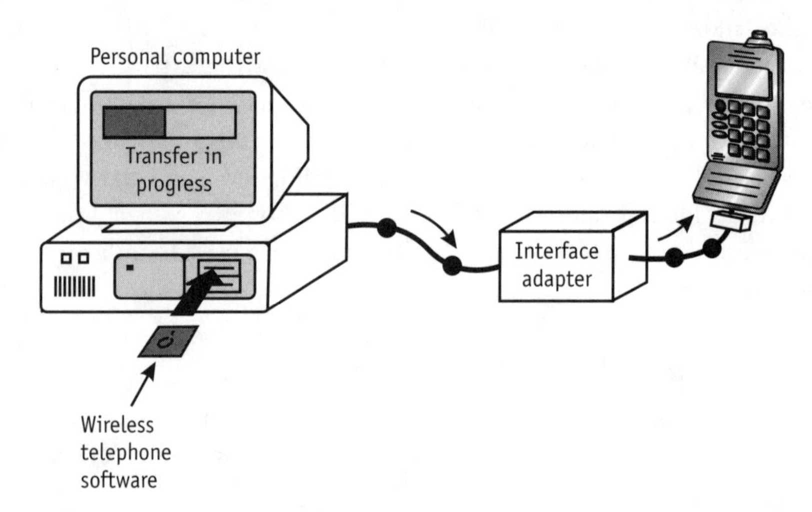

Figure 4.8 Software downloading.

Portable antennas commonly use 0- to 1-dB gain antennas, since people may turn the phone to many angles or leave the phone lying flat on the table. Figure 4.9 shows the different types of antenna gain.

4.6 EIA-553 AMPS telephone

A typical AMPS mobile phone is composed of several analog signal processing sections linked by a low complexity microprocessor (typically 8-bit) that is capable of processing approximately 0.5–1 MIP. The functional sections of an AMPS phone include an audio processing section, a modulator section, and an RF section. While the IS-136 system allows for dual AMPS and TDMA operation, the AMPS section of a TDMA phone typically shares the same digital processing sections. However, the following descriptions for functional sections still apply to the analog portion of a TDMA phone.

The transmit audio processes section prepares (enhances) the audio signal for FM transmission. The compressor section adjusts the dynamic range of the audio signal so that customers with different voice intensities (audio level) have approximately the same level of audio signal. By limiting the amount of audio level variance, the average amount of RF signal deviation is increased for low audio volume, thereby enhancing transmission performance. The pre-emphasis section provides additional gain for high frequencies. Since speech audio signals contain much of their energy at low frequencies, the pre-emphasis section increases the signal-to-noise ratio for the high-frequency components. The limiter section ensures that high levels of audio do not over-modulate the radio signal (splatter signal outside regulated bandwidth). The limiter has a unity gain until the audio level exceeds a level that would cause the deviation to exceed 5 kHz. The processed audio signal is then combined with the signaling SAT and STs as necessary. The modulator uses the RF synthesizer output with the complex audio signal (voice and STs) to produce an FM-modulated, 30-kHz-wide RF radio channel. The RF amplifier then boosts the signal for transmission; the RF amplifier output power is adjusted by the microprocessor control section in 4-dB steps.

A received AMPS radio channel signal is down-converted and amplified by the dual conversion RF receiver and amplifier section. The first

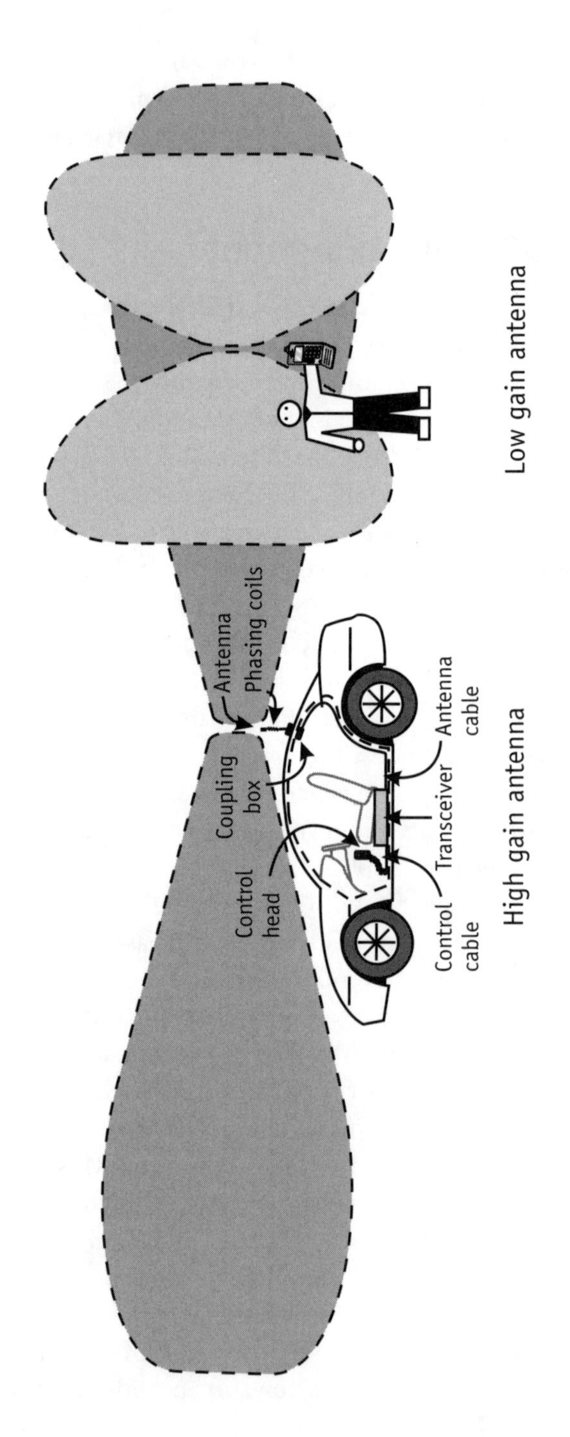

Figure 4.9 Antenna gain: (a) high-gain antenna and (b) low-gain antenna.

down-conversion mixer uses a variable frequency synthesizer to convert the incoming RF signal to a fixed *intermediate-frequency* (IF) signal. The second mixer converts the IF signal to a low-frequency signal (typically 455 kHz) for demodulation. The demodulated audio signal is supplied to audio filters so that the STs can be separated from the audio signal. The audio signal is then converted to its original analog (audio) form by processing it through the de-emphasis and expander sections, which reverse the effects of the pre-emphasis and compressor sections.

The microprocessor section of the AMPS mobile phone controls the overall operation. It receives commands from a keypad (or other control device), provides status indication to the display (or other alert device), and receives, processes, and transmits control commands to various functional assemblies in the mobile phone. Figure 4.10 shows a functional block diagram for an AMPS mobile phone.

4.7 IS-136 TDMA telephone

The complexity of an IS-136 TDMA mobile phone is greater than an AMPS mobile phone owing to the more advanced signal processing that is required. TDMA mobile phones use the same 30-kHz-wide radio channel; however, the baseband (digital audio) and broadband (RF) signals are very different. Typically, dual-mode IS-136 phones use the same circuitry to process the analog signal as the digital.

The transmit audio section samples sound pressure from the mobile phone's microphone into a 64-Kbps digital signal. The digital signal is then divided into 20 msec slots and sent to the speech coder. The speech coder compresses the 64 Kbps to a data rate of 7,950 bps. Subsequently, the channel coder adds error protection to some of the data bits using half-rate convolutional coding. This increases the data bit rate to 13 Kbps. The channel coder then adds control information (SACCH, FACCH, and DVCC), increasing the data rate to 16.2 Kbps. The data clock rate is increased to 48.6 Kbps (16.2 Kbps for each of three subscribers). The burst signal is supplied to a $\pi/4$ *differential quadrature phase shift keying* (DQPSK) modulator. Mixing the modulator with the output of the RF synthesizer produces a 30 kHz wide radio channel at the desired frequency (824–849 MHz). The RF amplifier boosts the signal for

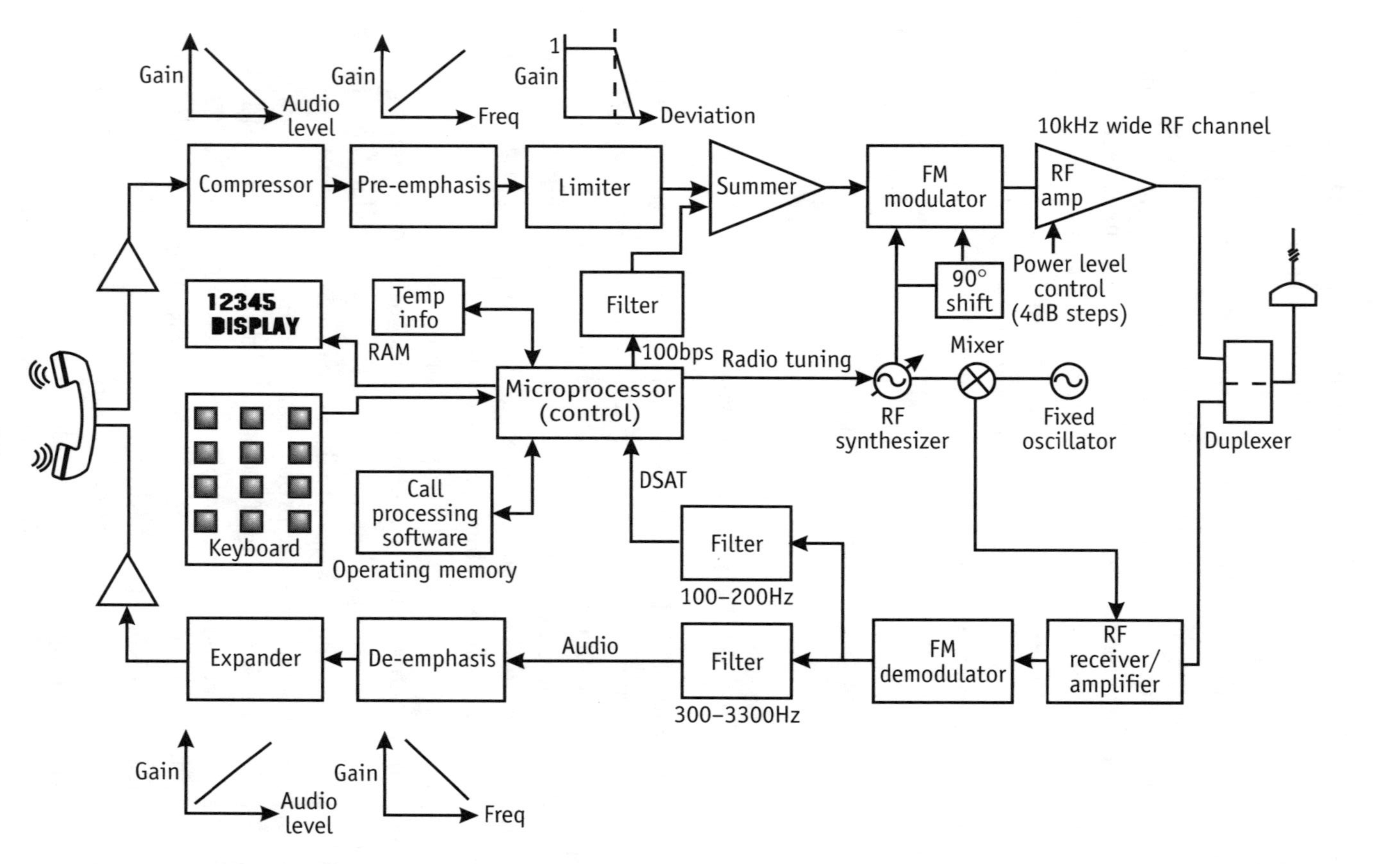

Figure 4.10 AMPS mobile phone block diagram.

transmission. The RF amplifier gain is adjusted by the microprocessor control section, which receives power level control signals from the base station.

A received TDMA signal is down-converted and amplified by the RF receiver and amplifier section. Because the incoming radio signal is related to the transmitted signal, a frequency synthesizer (variable frequency) signal produces the fixed frequency for down-conversion. The down-conversion mixer produces an IF signal that is either digitized and supplied to a $\pi/4$ DQPSK demodulator or down-converted by a second IF mixer to reduce the frequency even further (455–600 kHz). The demodulator may be adjusted by an RF equalizer, which helps adjust for distortions that may have occurred during transmission of the RF signal. The channel decoder extracts the data and control information and supplies the control information to the microprocessor and the speech data to the speech decoder. The speech decoder converts the data slots into a 64-Kbps PCM signal, which is then converted back to its original analog (audio) form.

The microprocessor section controls the overall operation of the mobile phone. It receives commands from a keypad (or other control device), provides status indication to the display (or other alert device), and receives, processes, and transmits control commands to various functional assemblies in the mobile phone.

4.8 Analog call processing

To correctly operate within a cellular system, a mobile phone must follow a specific sequence for processing signaling messages while attempting to make a call. This process, which is referred to as call processing, can be divided into four basic tasks: initialization, idle, access, and conversation. The mobile phone first completes the initialization task to obtain system parameters and register with the system. It then remains in the idle task awaiting new system information, pages, or a user command. When a call is initiated or is ready to be received, the mobile phone begins the system access task to secure a voice channel. After the system has assigned a voice channel, the mobile phone enters the conversation task.

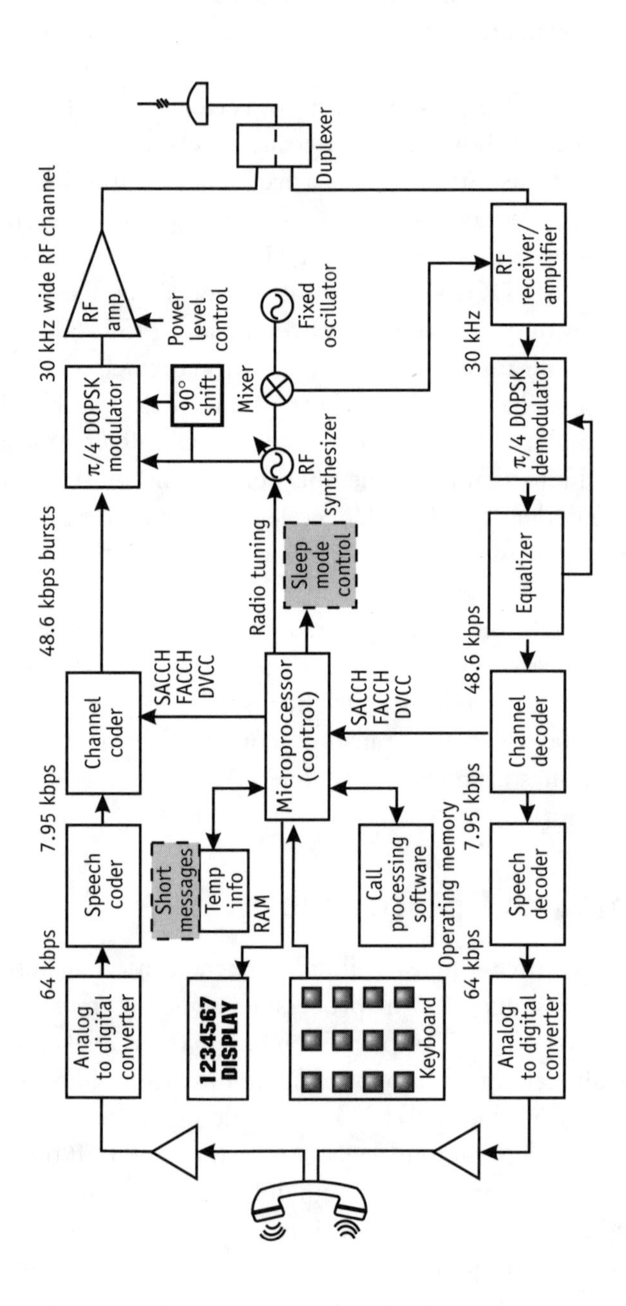

Figure 4.11 IS-136 TDMA mobile phone block diagram.

4.8.1 Initialization

When a mobile phone is first turned on, it must initialize its memory with information that is unique to the system in which it is operating. Cellular networks continuously send system overhead information on control channels to provide mobile phones with parameters for establishing communication and to inform customers of the system status. When the mobile phone is first turned on, it looks for dedicated control channels and locks onto the strongest signal. It then receives system parameters, such as SID and the number of PCHs.

4.8.2 Idle

After a mobile phone has initialized its memory, the phone must register with the system so that the phone's whereabouts are known. Subsequently, the phone continually monitors the control channel overhead information messages for changes in system information to obtain pages and to be ready for operation if the customer has dialed a number to initiate a call. It is necessary for the mobile phone to continuously monitor the system overhead messages to determine whether system access information has changed or whether a page has been sent to the phone indicating an incoming call. If the mobile phone has been paged, it will respond to the system on the strongest control channel and be ready to switch to a voice channel. If the customer has initiated a call, the mobile phone will send a call origination on the strongest control channel to indicate that it is attempting to access the system and make a call.

While monitoring a control channel during the idle mode, the mobile phone may move away from the serving cell site until the signal level of the control channel is below an acceptable level. When this happens, the mobile phone will scan for other control channels and tune to the strongest signal.

The phone must also inform the system of its whereabouts in order for the system to efficiently process calls. This is accomplished when the phone periodically registers with the system and enables the system to maintain accurate records of the phone status and location.

4.8.3 System access

System access is a random event in which a mobile phone attempts to gain the attention of a cellular system. A mobile phone attempts to access the system when the customer initiates a call or when the unit receives a page (or incoming call).

Before and during the access attempt message, the mobile phone will determine whether the system's control channel is busy by continuously monitoring the busy/idle bit status on the control channel. If the busy/idle bits indicate that the control channel is available, the mobile phone will begin to transmit an access request. The access attempt transmission contains a message that indicates which of these types of access is required (e.g. page response or call origination).

If the system determines from the access request information that service is authorized (that it is a valid customer), the system sends an IVCD message to assign a voice channel.

4.8.4 Conversation

After the mobile phone receives the IVCD message, it changes frequency to the new radio channel and begins voice communication. During conversation, the base station must continue to send control messages to the mobile phone that include power-level control and hand-off commands. Control messages are sent by blank and burst signaling. Blank and burst messages briefly replace voice information with signaling commands (for approximately 1/10th of a second).

To ensure that a reliable radio connection is maintained throughout a call, the SAT is continuously monitored to determine whether the radio signal is lost or whether an interfering mobile phone signal is being received in its place. If the mobile phone or base station does not detect the correct SAT tone for over five seconds, the radio link is determined to be broken and transmission must end.

During the call, the system may send hand-off commands. When the phone receives the hand-off command, it will mute the audio (to protect the customer from annoying sound transients), retune to the new channel, begin transmission, unmute the audio, and continue voice transmission.

When the call is complete, a release tone or message will be transferred. The release command has been modified from the AMPS system to include a DCCH locator message. This allows the phone to quickly find the DCCH if it is available.

4.9 IS-136 call processing

The call processing on the IS-136 system is more complex than that on the AMPS system. The IS-136 system has a new DCCH to coordinate access and provide advanced services and the DTC in addition to the analog voice channels of an AMPS system.

4.9.1 Scanning and locking

Scanning and locking is the process of finding the most suitable control channel, obtaining information about the system and acquiring the access parameters necessary to communicate with the system. If the channel is acceptable for gaining service, the phone will enter the DCCH camping state. If the channel is rejected (perhaps owing to poor signal strength), the phone will go back to the scanning and locking state in order to evaluate further channels.

When the mobile phone is first turned on, it will attempt to find service as quickly as possible. It does so using a process called selection. If the mobile phone fails to find a DCCH, it will begin to scan for an ACC. After finding an ACC, the phone may return to the DCCH environment if it finds a DCCH pointer on the ACC and if the prospective DCCH meets the camping criteria.

In addition, the phone may use intelligent roaming to find the correct frequency band (800-MHz or 1,900-MHz) to start scanning. Intelligent roaming takes into account the last band of operation in which the IS-136 phone was active. It also instructs the phone how to select a preferred channel in a mix of 800-and 1,900-MHz carriers. See Chapter 8 for more details.

4.9.2 Camping

The mobile phone is in the camping state when it is idle on a DCCH and waiting for pages from the system. During the camping state, IS-136 mobile phones perform many tasks, such as control channel reselections, in which the phone re-evaluates the DCCH environment in order to find the optimum channel on which to camp. In addition, the phone responds to internal and external messages to start call processing and registrations or receive short messages. Owing to the DCCH structure, the IS-136 phone is able to make neighbor control channel measurements during the phone sleep time so that no pages are missed.

An IS-136 mobile phone in the camping state must monitor its assigned paging slot to look for page messages from the base station. It must also check to determine whether any information has changed on the BCCHs. This check consists of a field on the paging slot that indicates when the broadcast information changes. The mobile phone only leaves its paging slot to read other slots in the DCCH when this field indicates new data.

A new function in IS-136 is MACA. MACA is a process of continually measuring the signal strength of new candidate control channels and reporting them back to the system. The system can use MACA for interference avoidance when it is setting up calls.

4.9.3 Access

A call from a mobile phone on the IS-136 system begins with an access to the system on the DCCH of the serving cell. The access includes an origination message containing the phone identification number (MIN), the serial number, and the called address. A call to a phone on the DCCH starts with a page from the system on the assigned paging slot and a page response from the phone. In either case, the system verifies the authenticity of the phone, the validity of the called address, and the availability of the network resources to handle the call prior to selecting an idle voice channel in the serving cell. In addition, the system first ensures that adequate signal strength is received from the phone.

One of the first steps to obtaining system access is running the authentication algorithm to verify the identity of the phone. To perform the authentication, the mobile phone uses information it has recently received from the system as part of the computation of the authentication information.

The voice channel assignment can be analog or digital depending on the phone capability, preference, and channel availability. An IVCD message is then sent to the phone. This message provides the phone with the selected voice channel to be used to start the conversation.

4.9.4 Conversation

The phone is in the conversation mode when it is tuned to a voice or DTC and is providing a voice path for a call. During conversation, the mobile phone continuously measures neighboring radio channels (both 800- and 1,900-MHz) and responds to system commands that may include hand-off to a new radio channel or time slot. When the conversation ends, the phone will retune to a DCCH and enter the camping state.

Key call processing events include receiving voice and control data, monitoring the status of the radio link, measuring radio channels that may be candidates for hand-off, changing channels when necessary, and disconnecting the channel when the call is completed.

4.9.5 FACCH and SACCH

During conversation mode, the phone continuously receives bursts of information from the base station. Each burst contains part of an SACCH message. After all parts of the SACCH message are received, the phone will process the complete SACCH control message. Each TDMA burst of information received also contains data. Most of this data is speech information. Occasionally, speech data is replaced with FACCH information. If the data is an FACCH message (e.g. hand-off message) the phone will immediately process the information and perform the appropriate action.

4.9.6 Mobile assisted hand-off (MAHO)

While operating on a DTC, the mobile phone can measure and report signal strengths received from other base stations that are hand-off candidates. The current base station uses the MAHO measurements in conjunction with base station measurements to generate hand-offs for phones being served on the DTCs. The MAHO process provides the system with mobile-station measurements with respect to surrounding cells and allows more accurate hand-offs to take place.

4.9.7 Hand-off

IS-136 allows for hand-offs between any combination of analog voice channels and DTCs at 800 MHz. The HCS algorithms may be taken into account in the hand-off process to provide coherent control and voice-channel borders within the system. Hand-offs can also occur between the hyperbands—from 1,900-MHz traffic channels to 800-MHz analog or digital channels and from 800-MHz DTCs to 1,900-MHz traffic channels.

4.9.8 Voice privacy

The IS-136 system provides a degree of cryptographic protection against eavesdropping in the phone-base station segment of the connection. The process of encryption involves using a data mask to encrypt the digital speech data and selected messages prior to transmission. Requests to activate/deactivate the voice privacy feature may be made during the call setup process, or when the phone is in the conversation state.

4.9.9 Discontinuous transmission

Discontinuous transmission (DTX) is another feature that may be used in the conversation mode, particularly in portable mobile phones. To conserve power, the mobile phone transmitter may be turned off during silent intervals in the conversation.

4.9.10 Call completion

When the call is completed, a release message is either received from the system or sent from the phone to terminate the call. In either case, the system message includes a DCCH pointer that enables a phone to return to a specific DCCH directly after a call with no need to rescan for a DCCH. In a properly configured network, the phone never has to randomly scan channels looking for service. The network tells the phone where to look for service.

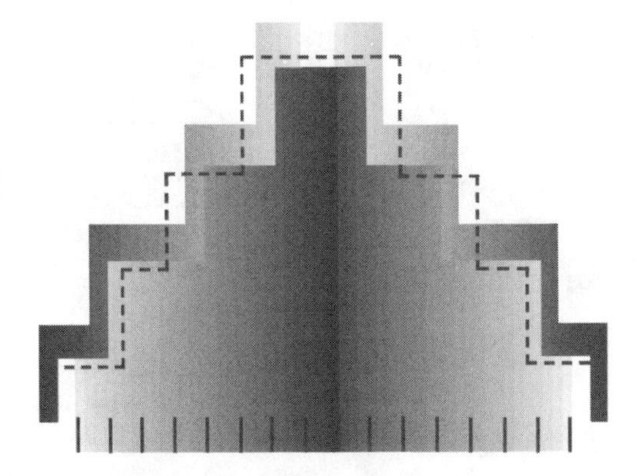

5

TDMA Networks

IS-136 TDMA NETWORKS integrate radio technology with network intelligence to provide advanced wireless services to a broad subscriber base. A wireless network is composed of mobile radios, cell sites, an MSC, a message center, voice mail systems, customer databases, and interconnections to various networks.

In the IS-136 network, the radio interface connects the customer to the cell site. The cell site converts the radio signal and transports the information to the MSC by way of a communications link. The MSC validates customer records in the various types of databases prior to authorizing service. The MSC connects the call to other mobile phones or to other networks. Figure 5.1 illustrates the fundamental interconnections in a wireless network.

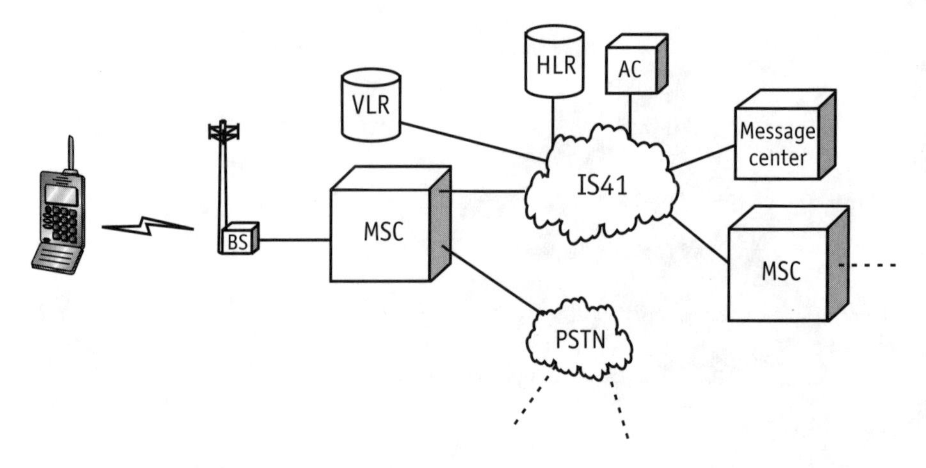

Figure 5.1 IS-136 wireless network.

5.1 Cell sites

Cell sites are composed of base station radio equipment and an antenna system. The base station radio equipment is typically located in a small environmentally secure building that is directly adjacent to the antenna tower. The base station holds the RF equipment (transceivers and antenna interface equipment), controllers, and power supplies. The antenna system consists of an antenna, a radio tower, cabling, and RF signal combining assemblies.

Base station transceivers have many of the same functional elements as a mobile phone. The radio transceiver section is divided into transmitter and receiver assemblies. The transmitter section converts the digital data representing control and voice information to RF for transmission to the mobile phone. The receiver section converts RF from the mobile phone to digital information to be routed to the MSC for further processing. The controller section commands insertion and extraction of signaling information.

Unlike the mobile phone, the transmit, receive, and control sections of a base station are grouped into equipment racks. For example, a single equipment rack may contain all of the digital radios and RF combiners. For analog cellular systems, one transceiver in each base station is dedicated to a control channel. In most digital wireless systems, control channels and voice channels are mixed on a single radio channel allowing a more efficient use of RF and equipment resources.

Figure 5.2 illustrates the components of a base station. Each assembly (equipment rack) typically contains multiple equipment assemblies for each RF channel. The assemblies include radio transmitters and receivers, power supplies, and antenna assemblies.

Because the IS-136 system was designed with the same radio channel bandwidth as AMPS, a digital radio channel can usually replace an analog channel to convert a radio channel from analog to digital service. To provide service for existing analog customers, each IS-136 base station in the 800-MHz band may have a number of radio channels for AMPS as well as digital radio channels. Figure 5.2 shows a typical IS-136 base station.

5.1.1 Transmitter section

The transmitter section contains modulator and RF power amplifier subassemblies. The modulator section converts the digital information into

Figure 5.2 Base station. (*Source:* Lucent Technologies.)

proportional phase shifts at the carrier frequency. A base station transmitter amplifier boosts the low-level modulated signals to typically 40–100W. The base station transmitter power level is normally fixed under normal operating circumstances since the power level determines the radio coverage boundaries of the cell site.

5.1.2 Receiver section

The receiver section contains a sensitive RF receiver and demodulator. The RF receiver boosts low-level RF signals and converts the frequency for input to the demodulator. The demodulator section converts the modulated RF signal to digital information. The receiver section also extracts control information and routes this to the base station controller. Since most base stations use dual antennas for diversity reception, the receivers must also be able to select or combine the strongest radio signals that are received on several antennas.

Cell sites typically request the signal strength information from several neighboring cell sites to determine the best candidate for hand-off. This requires each base station to have an RF receiver (sometimes called a locating receiver) that measures the RF signal level on any channel. The locating receiver provides the ability to measure any mobile phone's signal strength and channel for hand-off evaluation. IS-136 mobile phones that are operating on a digital radio channel can also report the RF signal level of various channels during the MAHO process. This information greatly improves the MSC's hand-off decisions.

5.1.3 Controller

The base station controller section consists of control signal routing and message processing. Controllers insert control channel signaling messages, set up voice channels, operate the radio location/scanning receiver, monitor equipment status, and report operational and failure status to the MSC. Typically, base station controllers have three parts: a base station, a radio communications controller, and an interface communications controller.

The controller coordinates the overall operation of the base station equipment based on commands received from the MSC. This includes continuously monitoring the operational status of the electronic assemblies in the base station. The base station also controls the radio communications links. This includes sending information on the control channel, coordination of radio channel assignment, and monitoring for link continuity during conversation.

5.1.4 Communications links

The base station is connected to the MSC by communications links. Communications links can use wire, microwave, or fiber optic links to transfer information. Communications links carry voice and control information between the MSC and the cell sites. To prevent potential outages of service in the event a communications link fails, alternate routes can be provided.

The typical link used for cell sites is a digital T1 line. The T1 line is a standard format for time-multiplexing digital communications channels in North America. Each T1 line can carry 24 channels. If a cell site uses a single T1 line, one channel is used for control, and the other 23 can be used to carry voice channels.

5.1.5 Antenna system

The antenna system consists of antennas, an RF combining section, and a receiver multicoupler. There are typically two receiver antennas and one transmitter antenna for each omnidirectional cell site. If the cell site has several sectors, there are two receiver antennas and one transmitter antenna per sector. Antennas vary in height from about 20 feet to more than 300 feet. Radio towers raise the height of antennas to provide greater area coverage. Because radio towers are expensive, it is often desirable for a single radio tower to share some systems—including paging systems and microwave links—with other antennas. To better integrate antennas into the environment, some antenna systems are camoflauged within their surrounding terrain or disguised as building structures.

The RF combiner allows several transmitter amplifiers to be connected to a single antenna. RF combiners are narrow bandpass filters that allow only one specified frequency to pass through. The filtering prohibits the output signals from one RF amplifier from leaking into another. The narrowband RF filter is typically a tuned cavity that allows only a narrow range of frequencies to pass. The dimensions of the cavity determine its filter frequency. To change the frequency, the chamber's dimension is changed. This is accomplished by manually turning a screw device or automatically changing it by a servo motor (autotune).

The receiver antennas are each connected to several receivers. This requires a receiver multicoupler to split the received signal. Unfortunately, splitting the received signal reduces its total radio energy to each receiver. To increase the receive level, the receiver multicoupler may contain low-noise RF preamplifiers to boost the received signals prior to the RF multicoupler splitter.

Two receive antennas are typically used to allow for diversity reception. Diversity reception reduces the effects of radio signal fading (called Rayleigh fading). Having two receiving antennas requires two receivers per channel to select the antenna that is receiving the stronger signal, or the receivers combine the incoming RF energy from both antennas in a process that produces a stronger signal.

If the cell site uses a microwave link to connect to the MSC, a microwave antenna may be mounted on the side of the tower. There may be several microwave links on a single cell site to allow connections to other cell sites.

5.1.6 Power supplies and backup energy sources

The electronic assemblies in the base station need regulated and filtered AC and DC voltage supplies along with air conditioning systems to cool the electronic equipment. Base stations are typically connected to the local power company. Since the operation of the equipment is critical for communications in emergency situations, batteries and generators are typically used to supply power to a base station when primary power is interrupted.

5.1.7 Cell site repeaters

Repeaters extend the range of a cell site coverage area by amplifying and redirecting the direction of the radio signal. Repeaters are used for large geographic rural areas and to cover hard-to-reach areas such as canyons and parking garages. Some repeaters operate by receiving, amplifying, and retransmiting the same frequency, while others receive, decode, and retransmit on a new frequency.

It is possible to use repeaters for the IS-136 system. However, to ensure that repeaters will work in the system, maximum delay time and linear radio signal amplification are needed. The propagation delays from repeaters may exceed the maximum delay offset that digital mobile phones can accommodate. In addition, repeaters that use class-C amplifiers introduce phase distortion and high or unusable *bit error rates* (BERs) for phase-modulated (digital) radio channels.

5.1.8 The mobile switching center (MSC)

The MSC processes requests for service from mobile phones and landline callers and routes calls between the base stations and the PSTN. MSC controllers receive dialed digits, create and interpret call processing tones, and connect calls via the switching system.

The MSC contains controllers, databases, a switching assembly, and power supplies. Communications controllers adapt voice signals to and from the communications links. Databases that hold customer information for home and visiting customers may be included. The switching assembly connects the communications links between base stations and the PSTN. Power supplies and backup energy sources power the equipment. Figure 5.3 shows a block diagram of an MSC.

5.1.8.1 Controllers

The MSC controllers coordinate base stations, MSC switching functions, and PSTN interconnections. The system controller (or subsections of the controller) creates and interprets commands between the MSC and the base stations, controls the MSC switch, validates customers requesting access, creates air time and PSTN billing records, and monitors for equipment failures. Communications controllers process and buffer voice and data information between the MSC, base stations, and the PSTN. Communications controllers combine or demultiplex the channels from high-speed communications links. The communications controllers also route call control commands (e.g., hang up and dialed digits) to the control assembly.

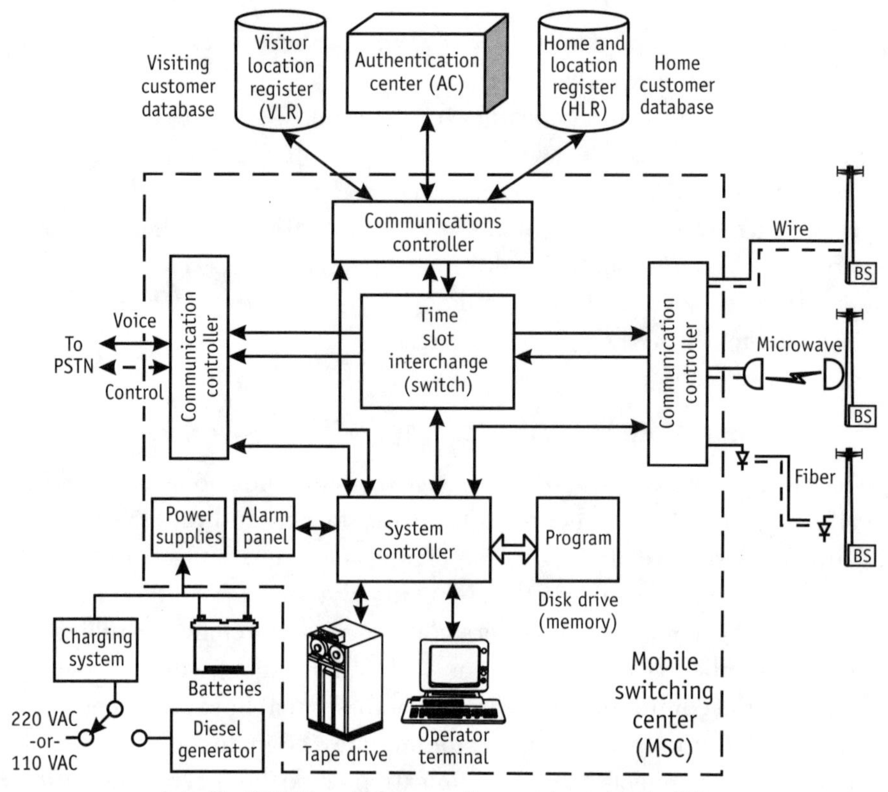

Figure 5.3 MSC block diagram.

In addition, controllers allow the connection of operator terminals that allow maintenance and administrative functions. The operator terminal is usually a computer monitor and keyboard dedicated to controlling equipment (e.g., configuring system parameters) and modifying the subscriber database. Operator terminal(s) may be located at the MSC or at other administrative facilities.

5.1.8.2 Switching assembly

The switching assembly connects base stations and the PSTN with either a physical connection (analog) or a logical path (digital). Early analog switches required a physical connection between switch paths. Today's digital cellular switches use digital communications links.

A switching assembly is a high-speed matrix memory storage and retrieval system that provides virtual connections between the base station voice channels and the PSTN voice channels. Time slots of voice channel information are input through demultiplexers, to be sequentially stored in the PCM data memory. Time slots that are stored in the PCM memory are retrieved and output through a multiplexer to the slots, which are routed to another network connection (such as a particular PSTN voice channel). By controlling the address location of the temporary memory locations, any input time slot (single channel) can be linked to an output time slot (different channel).

5.1.9 Communications links

Communications links connect the various parts of the cellular or PCS network. A single communications link has many 64-Kbps channels between the base station and switching system. In addition to the primary purpose of carrying voice traffic, one channel is typically reserved for control information that allows the MSC to communicate with the base station and the PSTN. To ensure reliability, alternate communications links can route channels through different network points if one or more communications links fail or is unavailable. Most communications links between cell sites and MSCs use a T1 *time division multiplex* (TDM) *pulse-coded modulation* (PCM) digital transmission system. Communications links between an MSC and the PSTN require a larger capacity communications link. High-capacity communications links are high-speed links that combine lower speed links. A single MSC may require hundreds of channels to be connected to the PSTN. Table 5.1 illustrates the commonly used communications links.

5.1.10 Home location register (HLR)

The customer database for a cellular or PCS system is contained in the *home location register* (HLR). The HLR contains each customer's MIN and ESN along with the customer's profile, which includes the selected long-distance carrier, calling restrictions, service fee charge rates, and other selected network features. Subscribers are sometimes allowed to change some of their feature preferences (such as call forwarding) in HLR by using their mobile phone. The MSC uses the HLR information to

Table 5.1
Communications Line Types

Number of Channels	Digital Signal Number	System Input	Bit Rate (Mbps)
1	DS0	1 PCM Voice	.064
24	DS1 (T1)	24 DS0	1.544
48	DS1C	2 DS1	3.152
96	DS2	4 DS1	6.312
672	DS3	28 DS1	44.736
4032	DS4	6 DS3	274.176

authorize system access and process individual call billing. The HLR can be part of the MSC or a separate processing platform dedicated to this critical database task.

5.1.11 Visitor location register (VLR)

When customers visit a new system (or roam), their customer information is retrieved from their home system's HLR and temporarily stored in a *visitor location register* (VLR). The VLR contains a copy of the customer's HLR information so that customers may experience regular service while roaming. Since the customer's information is temporarily stored in the VLR, it is unnecessary for the visited MSC to check with the visiting customer's HLR each time access is attempted. After a period of inactivity, or when the customer visits another system, the VLR information is erased.

5.1.12 Billing center

A separate database, called the billing center, keeps records on billing. The billing center receives individual call records from the HLR. The billing records are converted into a standard format to collect and process the information. The billing records are then transferred via tape or data link to a separate computer (typically *off-line*) to generate bills and maintain a billing history database.

5.1.13 Authentication center (AC)

The *authentication center* (AC), which processes information required to authenticate a mobile phone, is typically part of the HLR. During authentication, the AC processes information from the mobile phone and compares it to previously stored information. If the processed information matches, the mobile phone passes authentication and service is allowed.

5.1.14 Message center/teleservice server

The message center is a store-and-forward device that is used to support teleservices such as cellular messaging, OAA, OAP, or general UDP transport (for microbrowsers). The typical interfaces to a message center allow direct dial-in access (TAP), computer interface (TNPP) for e-mail delivery to phones, or operator input for paging services. The message center is informed of the success or failure of message delivery and can store and resend messages that were not correctly received. This means that even when an IS-136 phone is turned off or out of range, any messages sent to it will be delivered when the customer enters a coverage area or switches the phone on again.

5.1.15 Voice-mail systems

Voice-mail systems enable a caller to leave a verbal message when a called subscriber is not able to answer the phone. This is particularly useful in a wireless environment, since wireless subscribers do not always leave their phone on. For example, it may be inconvenient to answer the call during meetings or while driving.

In conjunction with call forwarding and other features, a voice-mail platform can be smoothly integrated with the wireless system so that calls can be sent directly to the voice-mail system when the phone is turned off or engaged in a call. Voice-mail retrieval generally uses DTMF signaling and can be easily accessed from a cellular phone. In addition, IS-136 systems provide a *message waiting indicator* (MWI) capability, which notifies the subscriber that a message is waiting in their voice mailbox.

5.1.16 Power supplies

The electronic assemblies in an MSC require power supplies. Power supplies consist of several AC and DC voltage supplies for switching equipment, subscriber databases, and cooling systems. An MSC also requires long-term backup energy supplies to operate the network system when primary power is interrupted. Backup power supplies are a combination of batteries and diesel generators. During normal operation, batteries are continually charged to be ready to supply backup power anytime. The batteries are directly connected to the network power supplies so that, when outside power is interrupted, they immediately and continuously power the system. After a short time without AC power, a backup generator automatically begins to supply power to the battery charger.

5.2 Public switched telephone network interconnection

Cellular or PCS networks are connected to the PSTN using several different types of connections. Early cellular systems connected to the PSTN through standard *plain old telephone service* (POTS) connections (dial tone). Unfortunately, POTS connections are inefficient and do not have advanced signaling capability. Today, wireless systems typically connect in the same way as a local telephone company connects to the PSTN *end office* (EO) connections. EO connections allow wireless systems to receive and send signaling messages (such as ANI and billing information) directly to the PSTN.

The PSTN is composed of various levels of switching and signaling. The lowest level of switching is called an EO class-5 switch. To interconnect EO switches, *tandem offices* (TOs) can cross-connect EO switches.

Control signaling in the PSTN is accomplished by sending control through a separate signaling network called *signaling system 7* (SS7). The SS7 network is composed of its own packet switches and switching facilities called *signaling transfer points* (STPs). STPs are controlled by *signaling control points* (SCPs). STPs are the telephone network switching points that route control messages to other switching points; SCPs are databases that allow messages—such as calling card information—to be processed as they pass through the network. Figure 5.4 shows an overview of the PSTN system.

Figure 5.4 shows two types of telephone networks: the *local exchange carrier* (LEC) and *interexchange carrier* (IXC) networks. LEC networks provide local telephone service to end users. IXC networks provide long-distance service. The U.S. government regulates how long-distance carriers and LEC carriers connect to end customers. IXCs typically connect to end customers through a *point-of-presence* (POP) connection. The POP connection is a location within a *local access and transport area* (LATA) that is part of the LEC's system.

The MSC must connect to the PSTN by way of a gateway connection. Various types of gateway connections can connect the MSC to the PSTN—type 1, type 2A, type 2B, and others. Type 1 connects the MSC with an EO as a standard POTS line (this is not typically used). Type 2 connections (there are various types of type 2 connections) link the MSC into a TO. In this case, the MSC appears to be a standard EO switching facility as part of the PSTN.

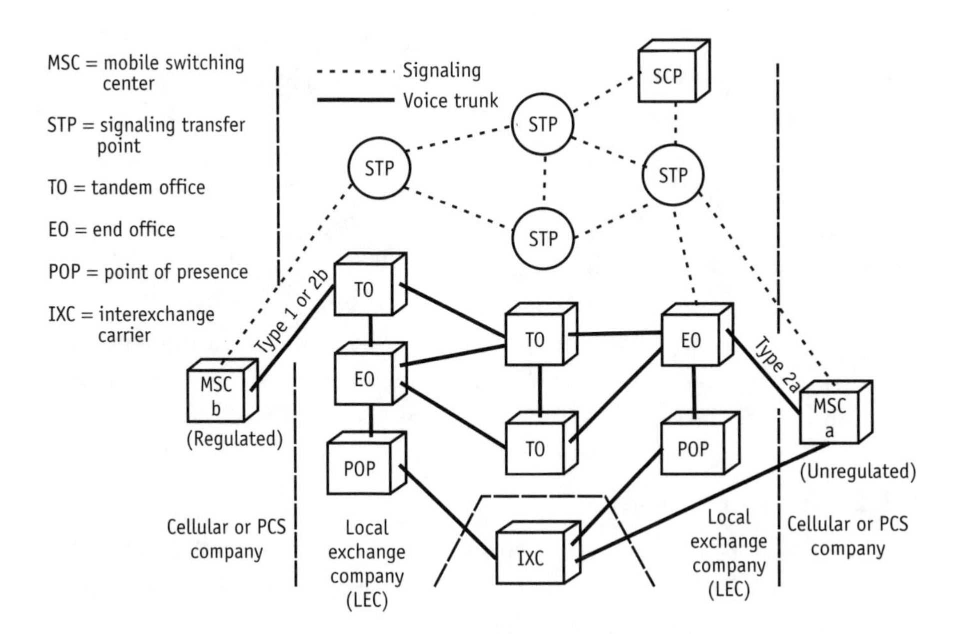

Figure 5.4 PSTN.

5.3 System interconnection

MSCs and other network nodes must be interconnected to allow customers to visit (roam into) different systems and to enable normal wireless operation across a distributed system. These system interconnections must allow validation of the visitor's identity, hand-off between systems, automatic delivery of calls, and advanced features operations.

To standardize the way in which wireless systems interconnect, the intersystem signaling standard IS-41 was created. IS-41 specifies how the wireless network nodes interconnect (x.25 or SS7) and how the messaging protocols are to be sent and received by each system.

5.3.1 Intersystem hand-off

Intersystem hand-off occurs when an MSC from one system transfers a call to the MSC of an adjacent system. Intersystem hand-off requires that the radio channel information be coordinated between adjacent systems. This is so that the serving system knows which radio channels are available on the adjacent system for hand-off.

Figure 5.5 illustrates intersystem hand-off between adjacent MSCs. The process begins when MSC-A determines that a hand-off is required. This is derived from the results of MAHO measurements and uplink signal strength measurements. MSC-A knows that a base station in an adjacent system is a potential candidate for hand-off and sends a message to MSC-B requesting it to verify the mobile phone's signal quality. If MSC-B allows the hand-off (there may not be any available channels), base station #1 will send a hand-off command that commands the mobile phone to change channels, and base station #2 will begin communicating with the mobile phone on the new channel. The voice path is then connected from the anchor (original) MSC-A to the new MSC-B, and the call continues.

5.3.2 Roamer validation

When customers visit IS-136 systems outside their home area, their identity is validated before they receive service from the visited system. Validation is necessary to limit fraudulent use of the service and to update the HLR about where to send calls.

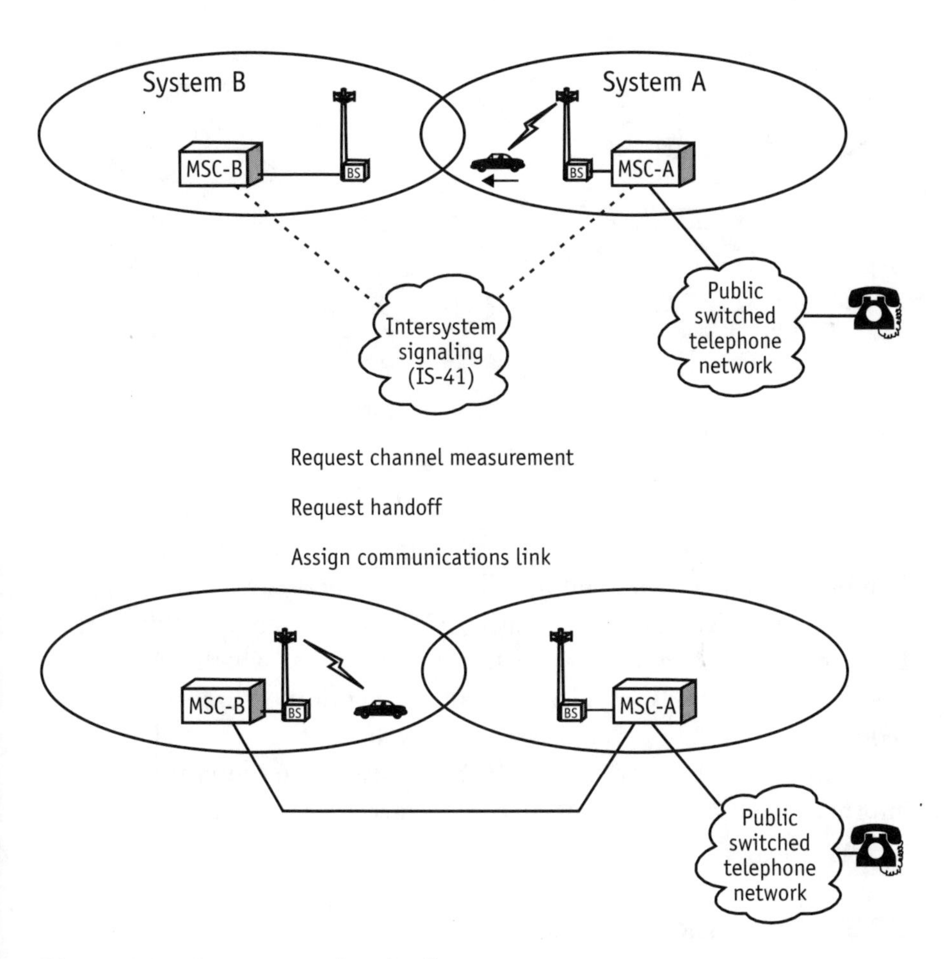

Figure 5.5 Intersystem hand-off.

Roamer validation uses various processes to verify a mobile phone's identity using registered subscriber information—usually by validating the MIN and ESN combination. Fraudulent activity has made it necessary to add further security procedures for additional validation called authentication. This is the processing and exchange of *shared secret data* (SSD) to confirm a mobile phone's true identity.

Figure 5.6 illustrates the typical roamer validation process. When a mobile phone initiates a call in a visited system, the visited system will receive the mobile phone's MIN and ESN when the mobile phone

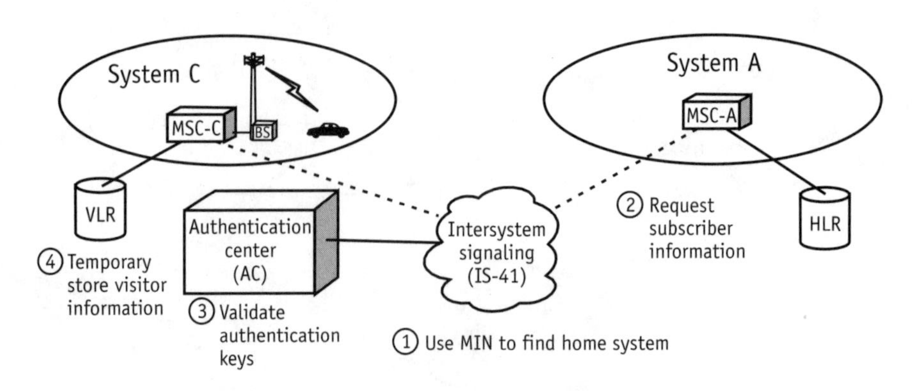

Figure 5.6 Roamer validation.

registers for service. The system will use the MIN to determine the location of its home system, and it will send a request to the home system for validation. The home system's AC/HLR will compare the MIN, ESN, and authentication information to determine whether it is valid customer. The HLR may pass some authentication information (if requested) so that the visited system can validate the customer without contacting the HLR again. The HLR will also store the visited system information so that calls can be routed directly to the new system in the future.

5.3.3 Automatic call delivery

The process of delivering a call to a mobile phone regardless of the system in which it is operating is called *automatic call delivery*. Automatic call delivery requires that the home system know the location of its customer's mobile phone at all times. The home system learns the location of a mobile phone during initial registration when the visited system provides the latest location information. This is accomplished using the *North American Cellular Network* (NACN).

5.3.4 Automatic roaming

The creation of a seamless cellular network across the continent plays an important role in simplifying the delivery of calls to customers—no matter where they are roaming. In North America, no one carrier can provide service to the entire population, so an industry-wide effort took place to provide seamless service across multiple carriers.

5.3.4.1 The North American Cellular Network (NACN)

The initial network of the NACN company, which was formed in 1991, served approximately 600,000 customers. Currently, over 100 cellular service providers in the United States, Canada, Mexico, Hong Kong, and New Zealand are part of the network, which serves more than 20 million customers.

The NACN is created by connecting cellular switches to a data network that uses SS7 signaling. Using the SS7 network, cellular switches communicate with one another with software written according to the IS-41 industry standard.

5.3.4.1.1 NACN Features The key features of the NACN are listed as follows.

- *Automatic call delivery:* The most well-known NACN feature is the automatic delivery of calls to customers, regardless of their location. As long as the customer is being served by a cellular system that is part of the network, the customer is automatically registered into the network and is ready to receive calls, just as he or she would do in their home system.
- *Features follow automatically:* Network features such as call waiting, call forwarding (immediate, busy, no answer), three-way calling, and call barring follow the customer seamlessly through the NACN.
- *Uniform feature access:* Customers are able to use the features to which they have subscribed anywhere in the network. A uniform method of controlling features has been implemented in all participating switches and includes standard codes for controlling features, such as the code *71 to activate call forwarding.

5.3.4.1.2 NACN Operation When a customer from an NACN market travels to another NACN market, the following events take place:

1. The customer turns on his or her phone. The phone immediately registers with the visited switch.

2. The visited switch looks at its NACN database and identifies the home of the customer.

3. The visited switch notifies the customer's home switch that the customer has registered in the visited market.

4. The home switch notes the location of the customer in its database and sends a complete subscriber profile to the visited switch. This profile includes the features to which the customer has subscribed as well as any restrictions placed on the customer in regard to calling areas.

5. The visited switch now sets up the customer in exactly the same way the customer is set up at home. Calls to the customer's cellular home phone number will be forwarded to the visited market cellular system.

This five-step process takes approximately two seconds. When the visited switch becomes aware that the customer has turned off his or her phone, the home switch is notified and the customer is "deregistered" in the visited market. The next time the customer turns on his or her phone, the registration process is repeated. Figure 5.7 depicts the autmoatic call delivery process.

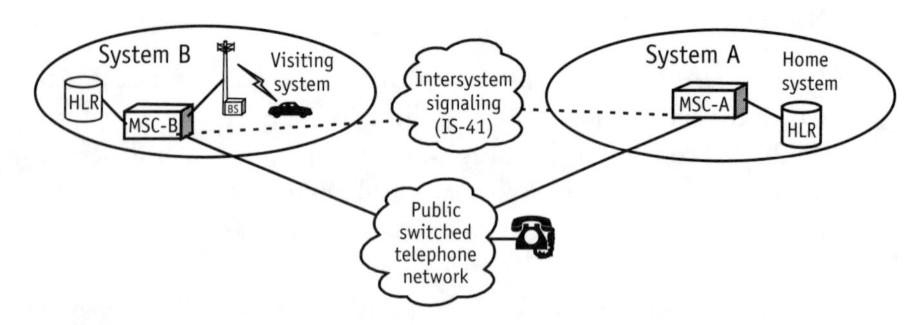

Figure 5.7 Automatic call delivery.

5.4 Private systems

Private systems allow customers to use their wireless phones to emulate desk phones in an office or campus environment. This process, which is known as WOS, enables customers to use an IS-136 phone as their desk phone in the office and as a normal wireless phone outside the office. The "office" could also be a warehouse, airport, hotel, hospital, or other workplace where employees are often unreachable at a desk or single location. There are two main facets to private systems, listed as follows.

1. In-building RF coverage, which may involve IS-136 technology engineered to the specific site;
2. PBX connection, which would typically be a T-span from the customer premises PBX to the nearest MSC.

In addition to operating a private system in single locations, IS-136 private systems can be configured at multiple customer sites across the country (or world) and provide customers with the same features they receive in their regular office location—for example, flat-rate billing or a variety of PBX functions.

Private systems do not have to be configured to serve private customers exclusively. In fact, in many cases, the enhanced traffic capability and coverage of the private system is used to serve regular wireless subscribers who happen to be in the vicinity of the private system. For example, in a hospital, the nursing staff may function as private-system customers with a special billing rate, whereas the visitors would be treated as regular customers. IS-136 uses semiprivate designations and intelligence in the network to support such flexibility.

Private systems can be supported by an existing macrocell in a wireless system or by microcells placed at strategic locations around the area to be covered. In either case, the radios are controlled from the MSC in the wireless system. Autonomous systems can also be used to support private systems. Autonomous systems are self-contained radios and switching equipment that perform self-RF engineering to adapt to the existing wireless system and that may also include PBX functionality. Autonomous systems can coexist with the regular wireless system and provide a cost-effective solution to private systems and WOS. Phones using IS-136 protocols and advanced reselection techniques are able to access the autonomous systems and, thus, offer the same features and services as the regular IS-136 systems.

5.4.1 Tools for private-system engineering

IS-136 provides a number of tools to facilitate the creation and operation of private systems. These tools allow the system designer to designate cells as part of a private system, to inform phones using broadcast identifiers which particular private system company is being supported on that cell, and to control the radio environment to allow phones to give preference to the cells providing coverage for the private system. The following IS-136 features can be used to achieve these goals:

- *Network type:* A broadcast identifier that can be used to specify a cell as public, private, residential, or a mixture of all three.
- *PSIDs:* Broadcast PSIDs are used to mark a cell or collection of cells geographically to enable location-based services such as alpha tags (system name).
- *HCSs:* An indication on a cell's broadcast neighbor list of a preference to each neighbor so that the coverage and influence of a private system can be carefully controlled. Different algorithms are used for reselection or hand-off depending on the neighbor's attribute of regular, preferred, or nonpreferred. These features are used to manage traffic more effectively and to allow greater control of the radio environment.
- *Registration:* New registration types have been developed so that phones that recognize a PSID or SID change (meaning that they have entered or left the geographical location of their own private system) will inform the cellular network of this change of system.

5.4.2 A private-system example

The private-system example shown in Figure 5.8 helps explain the use of the IS-136 tools currently available for setting up private systems (in this case, a small business).

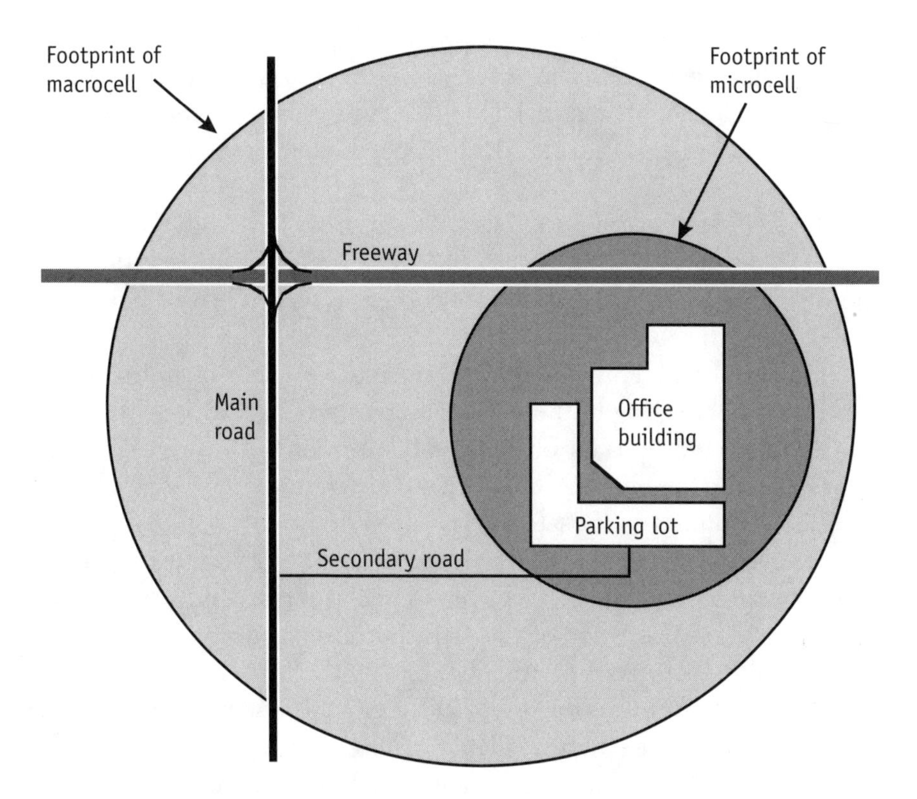

Figure 5.8 Private-system microcell configuration.

5.4.3 Private-system microcell configuration

In Figure 5.8, a microcell may be used to provide coverage in the building if the signal strength from the macrocell was insufficient inside the building or if the estimated traffic load from all the wireless phones in the office building would place too great a demand on the umbrella macrocell.

Hierarchical cell structures may be employed to configure the hand-off and reselection around the office building. As mentioned in Chapter 3, in some instances, the strongest available channel is not necessarily the most suitable. This would be the case if microcells were used to increase capacity at a certain location that was also covered by an umbrella macrocell. Without HCS, phones would not be able to camp on the microcells since the "standard" AMPS algorithms dictate that a phone select the strongest channel. HCS allows phones to camp on weaker channels that meet a sufficient signal strength criteria when appropriately designated. Hence, the microcell in this example may be marked as a preferred cell to force traffic onto the microcell while in the location of the office.

The microcell serving the private system will broadcast a network type of *public and private* or *semiprivate*. This means that the microcell will support public or regular wireless customers as well as private customers who work at this location.

Subsequently, a PSID will be chosen and broadcast from the microcell. Phones belonging to the private system will have that PSID, along with an associated alpha tag or location ID relevant to the private system, preprogrammed in memory either manually or using the OAA teleservice. When phones belonging to the private system come within the coverage area (defined using the HCS), the broadcast PSID will be matched to the PSID stored in the phone. The phone will perform a new system registration, and, on reception of a registration accept, the alpha tag (system banner) will be displayed on the phone.

The alpha tag informs the user that the enhanced services provided by the private system are now available. These services can include four-digit dialing, private numbering plans, and lower airtime rates. Phones that do not belong to the private system will still receive wireless service from that cell but will not register, display the private system alpha tag, or have access to the enhanced services offered to private system customers.

When the phone leaves the private system coverage area, the phone reregisters with the public umbrella macrocell and removes the alpha tag from the display to inform the user that the special services are no longer available. The display will then return to the alpha tag of the regular wireless carrier. Figure 5.9 shows a private system with hierarchical cell structures.

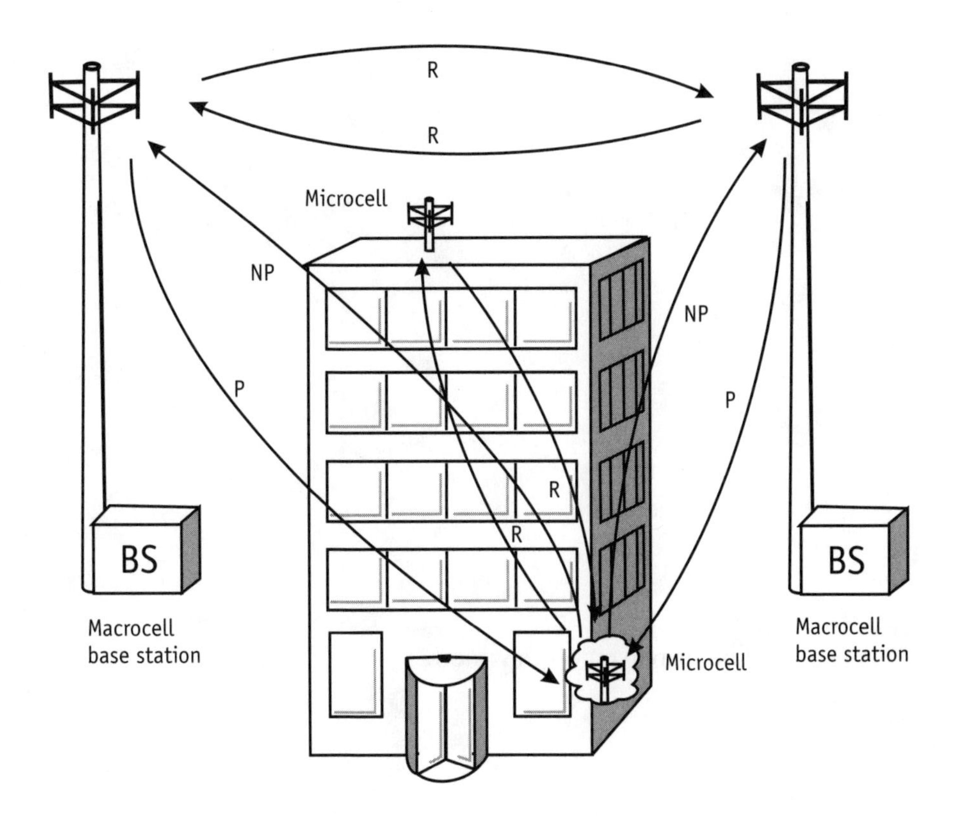

Figure 5.9 Private system with hierarchical cell structures.

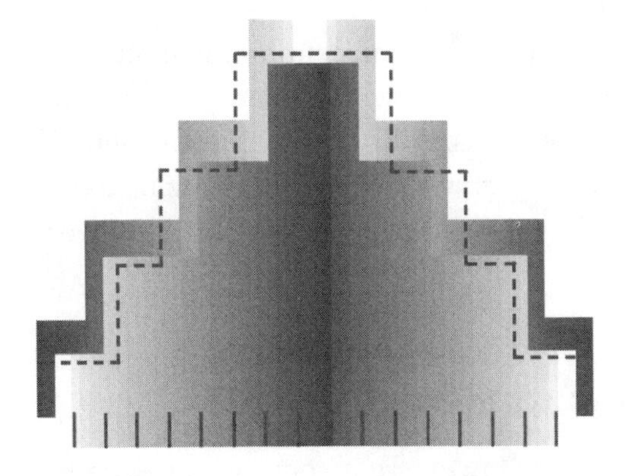

6 Testing

Testing involves verification of operation and tolerances associated with mobile phones and system equipment. While the tests for analog and digital signals measure different parameters, the resultant performance attributes (clarity, robustness, etc.) that are affected by these parameters are similar. Test requirements for IS-136 mobile phones are located in IS-137-A, TDMA Cellular/PCS—Radio Interface—Minimum Performance Standard for Mobile Stations. The test requirements for IS-136 system equipment are located in IS-138, TDMA Cellular/PCS—Radio Interface—Minimum Performance Standard for Base Stations.

6.1 Analog versus digital testing

Testing dual-mode mobile phones is different than testing analog phones. While a single piece of test equipment may be used to test both the analog and digital functions of a dual-mode phone, the instrument will often use completely separate subsystems for each mode. In general, most of the analog measurement is accomplished using traditional audio and RF measurements, but the digital measurement is done by digitizing a portion of the signal and using digital signal processing to extract parametric information.

6.2 Transmitter measurements

Transmitter tests typically involve RF power, frequency, and modulation quality measurements. Such tests determine whether the mobile phone can correctly operate within the bandwidth, frequency, and RF power constraints placed on it by the industry's operating standards.

6.2.1 Transmitter RF frequency

The frequency of an IS-136 mobile phone in AMPS mode is simple to measure. To do so, simply enable the transmitter with or without any input audio signal and use a frequency counter to average the frequency over a second or so. Measuring the frequency and phase of the IS-136 TDMA channel is different. The modulation is constantly changing the phase even without an input signal, and, more significantly, that phase change is always advancing by, on average, one-eighth of a cycle per symbol. In addition, the frequency of all dual-mode phones and most modern analog phones is locked to the frequency of the incoming (received) signal. To measure the frequency and phase accuracy of a mobile phone, the frequency of the forward channel must be accurately referenced and the received signal—when in digital mode—must be measured by digitizing an entire burst, decoding the data, and calculating the effective frequency mathematically.

6.2.2 Power measurements

Because the pulsed transmissions require a different type of power meter, a standard power meter should not be used to measure the RF power on a DTC. Most power meters detect peak power.

The best way to test the power of a digital phone is to use a high-speed power analyzer. These instruments use special power meter sensors to measure the instantaneous power level and to plot it on an oscilloscope-type display. They can then be used to plot the RF power over a complete burst or average between two particular symbols. Unfortunately, these meters tend to be expensive. Most newer single-box test systems measure the burst power by digitizing an entire RF signal burst and using digital signal processing to calculate the power level.

6.2.3 Modulation quality

In order to pass information on a carrier, it must be modulated. To measure the quality of this modulation, different tests are performed in analog and digital modes. In analog mode, quantities such as transmit deviation in both voice and data modes, residual deviation with no audio input, SAT deviation, distortion, *signal to noise and distortion* (SINAD), compression, and limiting are all measured using a basic FM receiver. When the radio carrier is DQPSK-modulated, as on a digital channel, some new parameters are measured. Quantities such as vector error measurements over one full burst and over the first ten symbols of ten bursts, along with origin offset allows us to determine the accuracy of the digital modulation. Once again, these digital quantities are measured by digitizing an entire burst and calculating them mathematically.

6.2.4 Adjacent and alternate channel power

When the radio carrier is phase-modulated, it will produce small amounts of radio energy outside its designated radio channel. When this energy falls in the channel that is directly above or below the designated channel, it is called adjacent channel interference. When radio energy spills into the radio channel that is two channels above or below, it is called alternate channel interference. To measure alternate and adjacent channel interference, the mobile phone is set up at full transmitter power and the radio energy is measured outside the channel bandwidth. Figure 6.1 illustrates the RF spectral mask for alternate and adjacent channel interference.

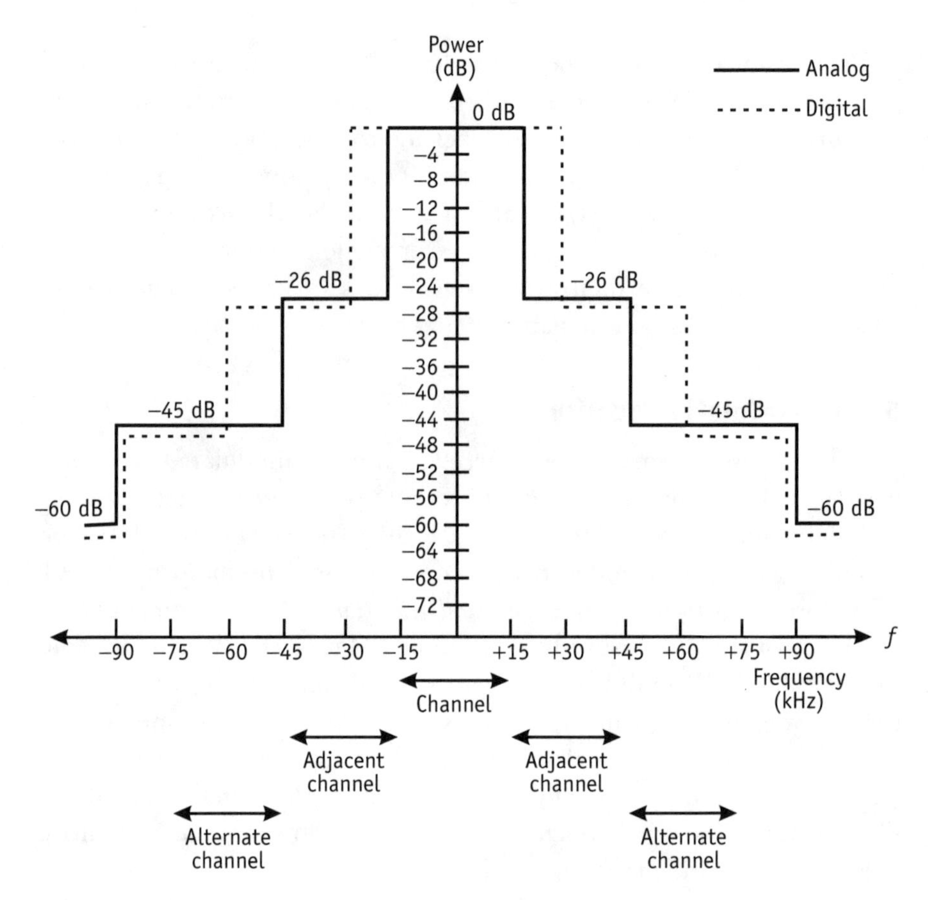

Figure 6.1 IS-136 RF spectral mask.

6.3 Receiver measurements

6.3.1 RF sensitivity

The standard test for analog receivers is the SINAD test, which involves providing an RF test signal that is modulated with a fixed tone; a signal distortion analyzer monitors the difference between the transmitted audio signal and the received audio signal. The RF signal strength is continually lowered until the distortion level exceeds the distortion tolerance. When this occurs, the RF signal level determines the sensitivity of the receiver.

Standard SINAD measurements cannot be performed on an IS-136 digital receiver because the speech coder only processes digital information. However, the RF sensitivity test for a digital receiver is not very complicated. Instead of using the audio distortion level, a digital signal is supplied to the modulator and the digital output is compared bit for bit to determine a BER. The RF signal source is continually decreased until the BER limit is exceeded. This is the RF sensitivity level for the digital receiver.

6.3.2 Demodulated signal quality

For analog mode, the quality of the demodulated audio signal can be measured by modulating the carrier with various sinewave test tones and measuring distortion, SINAD, signal-to-noise ratio, expansion ratio, frequency response, and audio level. Using a test "head," these measurements can also be made on the acoustical signal coming from the phone.

However, these measurements become more complicated in digital mode. When the phone processes voice signals, it passes them through a vocoder to reduce the effective bit rate. The vocoder is optimized to process speech only and introduces a large amount of distortion when supplied with nonspeech-like signals. Worse still, most modern phones have DSP-based noise cancellation algorithms that see these test signals as noise and actively try to remove them!

To measure the performance of the audio section of digital phones requires the use of a standard DAT tape to supply a known segment of speech signal to both the phone and channel 1 of a two-channel FFT analyzer; the resulting audio from the receiver is applied to channel 2. The difference is then analyzed for frequency response and other parameters using the analyzer's internal mathematical functions.

6.3.3 Cochannel rejection

Cochannel rejection is the ability of a mobile phone's receiver to differentiate between a desired radio signal and a weak interfering signal that is operating on the same frequency. To measure cochannel interference, a reference signal is applied to the receiver with a known digital pattern, and an interfering RF signal is applied at the same frequency with a different digital pattern. Similar to the receiver sensitivity test, the digital output of the mobile radio is compared to the original (desired) digital input signal. The RF signal level of the interfering signal is continually increased until the BER exceeds a maximum limit. The difference between the desired RF signal level and the RF signal level of the cochannel interference signal is the cochannel rejection ratio. Figure 6.2 shows how cochannel rejection is measured.

6.3.4 Adjacent channel rejection

Adjacent channel rejection is the ability of a mobile phone to differentiate between a strong RF signal on an adjacent radio channel and the desired signal on the current channel. The test procedure to measure adjacent channel rejection is similar to the cochannel rejection measurement radio. A reference signal is applied to the receiver with a known digital pattern, and an interfering RF signal is applied at a radio channel above or below with a different digital pattern. To determine the maximum level of adjacent channel interference that can be tolerated, the RF signal level of the adjacent channel interference signal is increased until the BER of the digital output exceeds a maximum limit. The difference between the desired RF signal level and the RF signal level of the adjacent channel

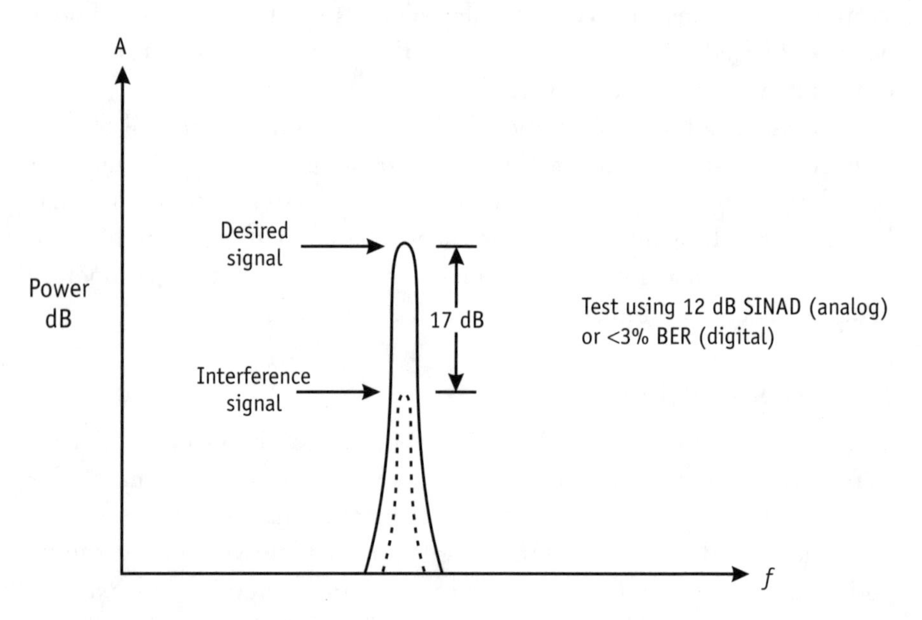

Figure 6.2 Cochannel rejection measurements.

interference signal is the adjacent channel interference rejection ratio. Figure 6.3 shows how to test for adjacent channel rejection.

6.3.5 Intermodulation rejection

Intermodulation distortion occurs when undesired frequency components are created in the receiver section as a result of the mixing of two or more RF interference signals. When these signals mix in the receiver, they may produce a signal at the frequency that is supplied to the modulator. To test intermodulation rejection, a desired reference signal is applied to the receiver in the presence of radio signals that have a sum or difference in frequency that may produce an unwanted receiver frequency. To determine the amount of intermodulation rejection that can be tolerated, the RF signal level of the interference signals are increased until the BER of the digital output exceeds a maximum limit. The

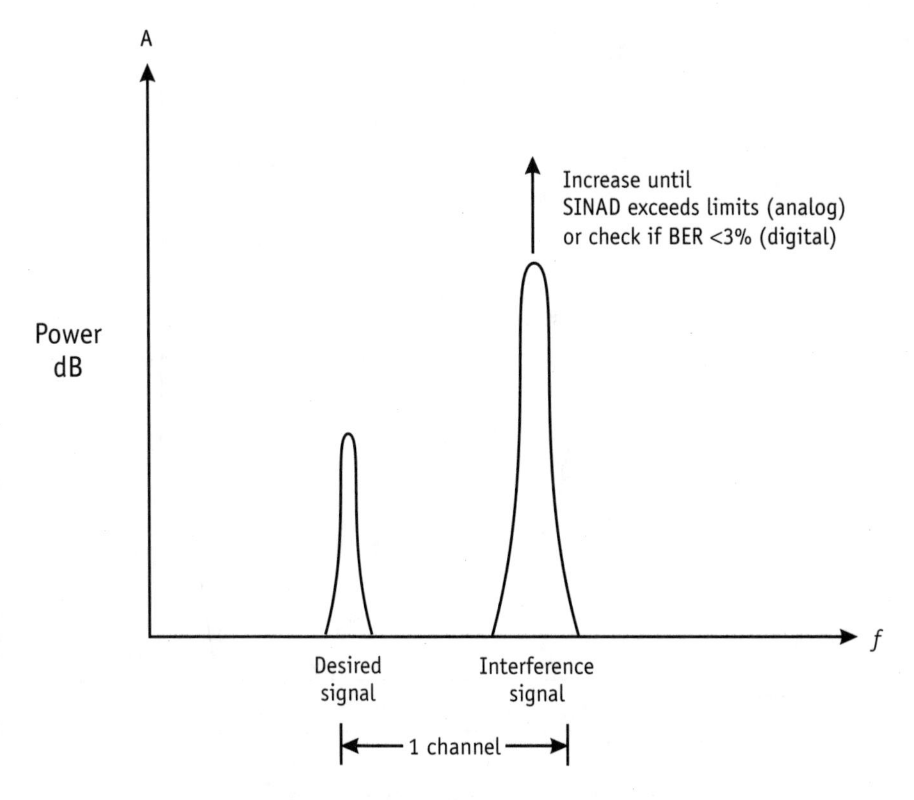

Figure 6.3 Adjacent channel interference rejection measurements.

difference between the desired RF signal level and the RF signal level of the interference signals is the intermodulation rejection ratio. Figure 6.4 shows how to test for intermodulation rejection.

6.3.6 Multipath fading

Multipath signals result when a desired RF signal is received along with a reflection of a signal that is delayed in time. This happens on a regular basis, because reflected signals bounce and travel over different paths, delaying the RF signal's arrival at the receiver. The signal delays cause amplitude and phase distortions in the desired waveform, and this results in errors. IS-136 radio receivers are designed to compensate for multipath signal distortion through the use of an equalizer. Equalization compensates for the varying radio channel conditions by estimating the likely distortions due to a delayed signal and subtracting these distortions from the desired signal.

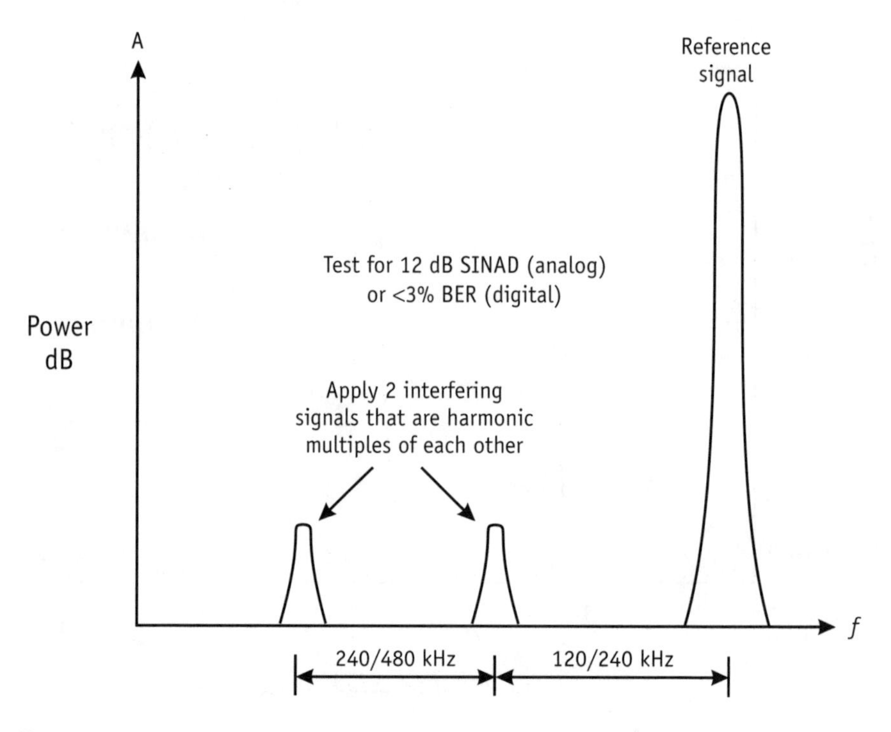

Figure 6.4 Intermodulation rejection measurements.

To test for resistance to multipath signals, a test generator is used to create a desired radio signal and a delayed radio signal. For a fixed delay (e.g., 20 μsec), the RF signal level of the interfering level is increased until the BER of the received digital signal exceeds its limit. Figure 6.5 shows a test signal generator that is capable of creating interfering radio channels.

6.3.7 MAHO measurement

In order to actively participate in the MAHO functionality of the system, the phone must be able to measure the RSSI and BER of the channel that it is currently on and the RSSI of up to 24 others. To do this, the phone is supported on a call using a cellular test box and given a neighbor list of up to 24 channels that are then supplied using a generator similar to that shown in Figure 6.6. The measurement is then requested from the phone, and the results are compared with the supplied level.

Figure 6.5 Test signal generator. (*Source:* Marconi.)

Figure 6.6 Multichannel RF signal generator. (*Source:* Berkeley Varitronics.)

6.4 Mobile telephone testing

6.4.1 Testing of mobile phones

The testing of mobile phones may be broken down into four categories: laboratory testing, field testing, parametric testing, and environmental testing.

6.4.2 Laboratory testing

During this stage of the testing, all aspects of the phone's operation with real infrastructure are tested whilst hard-wired to base station(s). The idea here is to ensure that the protocol handling of the phone and its user features are fully operational with a known RF environment. This testing can start when the phone is not yet complete and is usually done by a combined team from the customer and the manufacturer. The completion of this testing may require software changes in the phone when problems are discovered. The testing completed here ensures that the phone conforms to IS-136(a) and to any special requirements the vendor may have. Some of the items tested include selection, reselection, intelligent roaming, call origination, paging, SMS, hand-off, CNI, MWI, and MAHO.

6.4.3 Field testing

Once the operation of the phone has been verified in the lab, the unit is taken out in a test vehicle and a number of the tests are repeated to ensure that they work in a real-world environment. Carriers and vendors have set up drive test systems in which this testing may be achieved without endangering paying customers on the live system.

6.4.4 Parametric testing

Once the phone reaches commercial status, and the hardware platform and software are finalized, RF parametric testing may be completed. It is usually completed in a shielded RF lab using dedicated test equipment—rather than the real system—to provide a simulated infrastructure environment.

6.4.5 Environmental testing

In order to ascertain the operation of the phone in adverse conditions, a subset of the RF parametric testing is repeated at various temperatures and humidity levels and before and after shaking, dropping, blowing dust at, and shocking the phone to levels specified in IS-137 and the carriers' own specifications.

6.4.6 Test interface adapter

To enable testing of mobile phones, the requirements for a test interface adapter were specified in IS-137 and its predecessor, IS-57. However, after the first generation of phones, manufacturers generally dropped this item from their phones, providing only audio input and output and an RF jack or adaptor. Without such a test interface adapter, it is impractical to rely on accessing any sort of test mode, as different test instructions are required for different manufacturers of telephones. Accordingly, as much as possible, testing is done with the phone operating much as it would be in the customers' hands.

6.4.7 Test equipment

Factory testing confirms the tolerance of almost all electronic circuits in a radio, while field testing is used to validate that a mobile phone is operational. Production test equipment is typically integrated into the manufacturing facilities computer system. This means that production equipment allows external control. The manufacturing company writes test software to control the automatic testing and adjustment of mobile radios. As manufacturers are dealing with only their own brand of radio, they have access to special test commands that allow access to internal segments of the unit and can test individual parts (e.g., the front end) separately for performance. Figure 6.7 shows a production tester.

Precision test equipment is expensive and bulky. This has led to low-cost portable field test units. Field test equipment is typically designed for

Figure 6.7 Production tester. (*Source:* Hewlett Packard.)

specific technologies (such as IS-136) and has the capability to operate on 12V. Figure 6.8 shows a IS-136 TDMA tester.

6.5 Network testing

6.5.1 Radio propagation testing

System verification includes the determination of RF coverage quality and system operation. Signal quality is tested for individual cells first; then adjacent cells are measured to determine system performance. System operation is verified by measuring the signal quality level at hand-off, blockage performance, and the number of dropped calls.

RF coverage area verification ensures that a minimum coverage area percentage has been satisfied. The received signal strength and

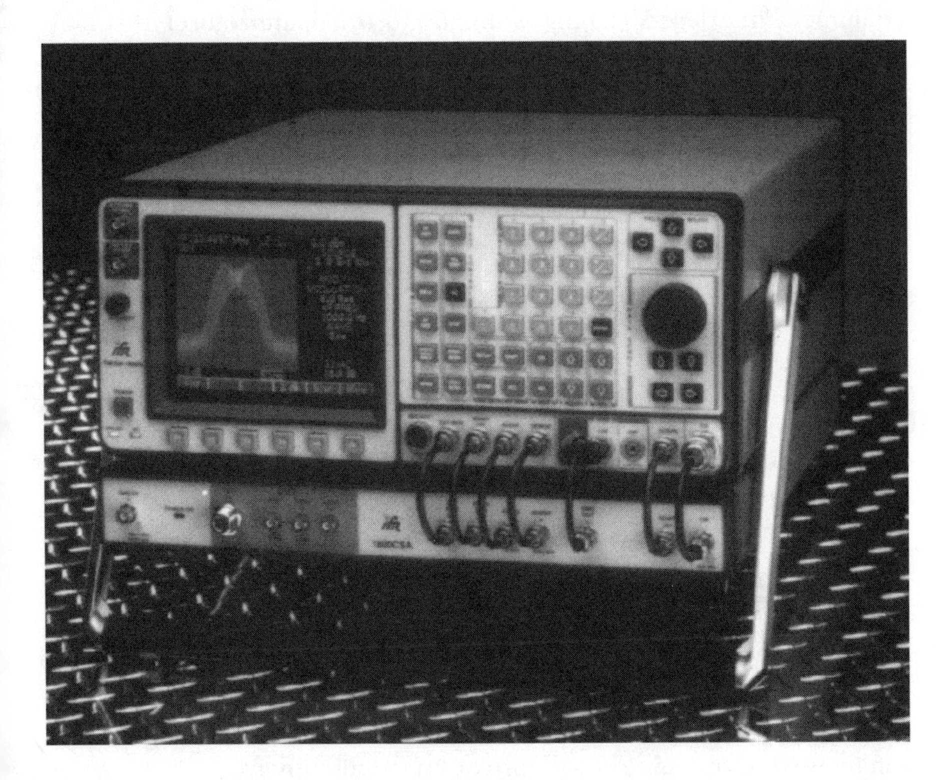

Figure 6.8 Field tester. (*Source:* IFR.)

interference levels are measured to locate holes where the signal level falls below an acceptable level due to terrain and obstructions. These areas may require another cell site or microcell or may be filled in with repeaters that amplify the existing signal and focus the energy into dead spots (e.g., a parking garage). However, IS-136 operation is best achieved using a microcell approach, rather than repeaters, as the cell parameters can be set individually for the area being covered.

System operation can be verified by recording the input levels received by a test mobile station when hand-off and access occur. Blocking probability can be calculated from the number of access attempts rejected by the system.

6.5.2 Network equipment testing

Network equipment testing is part of the *operations, administration, and maintenance* (OA&M) portion of the network. Most IS-136 network equipment interface signaling is unique to the manufacturer that constructed the equipment. These manufacturers provide test documentation and any interface adapters that may be required to independently test individual pieces of network equipment.

Cellular and PCS networks are designed for reliability. Many assemblies and communications links are redundant so that they can be placed out of service in the event of a failure. Networks have many automatic test systems to determine whether a piece of equipment has failed. If the test system determines a failure or out-of-tolerance condition, the equipment may be disabled and tested.

Networks typically have test systems that can use parts of the network for loopback testing of assemblies throughout the network. Figure 6.9 shows a sample system loopback test system. In path 1, the network sends a test signal to the base and routes the signal back to the system switch through the communications card. If the signal is successfully returned, the communications path and part of the communications card are verified as operational. For path 2, the test signal is directed to continue to the transmitter where it is sampled and sent back through the receiver. If the signal is successfully returned, the transmitter and receiver sections are also verified as operational. Various other network test configurations can be used to test and reconfigure network equipment.

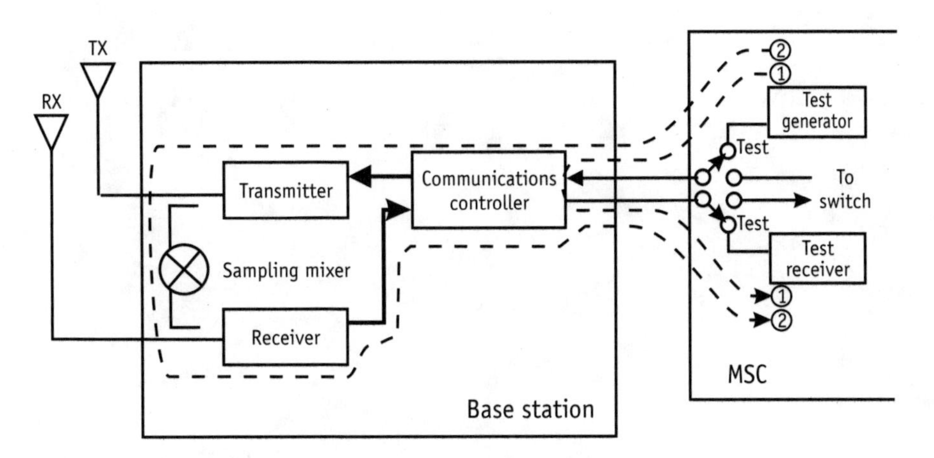

Figure 6.9 System loopback testing.

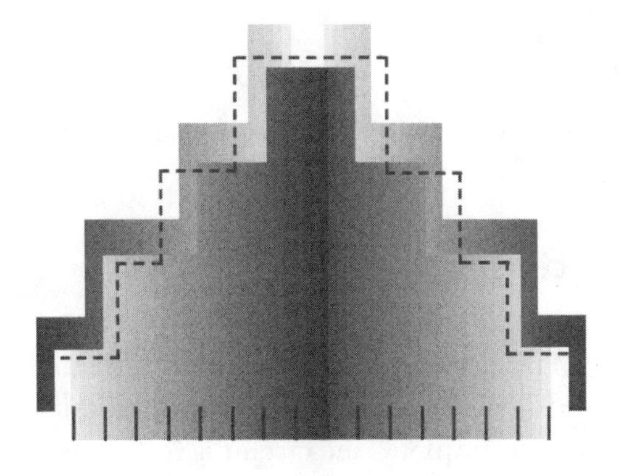

7

Data Services in IS-136

7.1 Overview

Many new data services are available through the IS-136 system. This chapter provides a technical description of these services along with a look at some other data features including packet, direct IP access, and phone-based microbrowsers.

Wireless data capability is becoming an increasingly important part of the wireless service provider's portfolio with the increased need for connectivity away from a user's desk. The adoption of handheld computing devices like palmtops and hand-held devices (Win CE) and the more widespread use of laptop computers has also fueled that need. In addition, the introduction of the SMS into the cellular and PCS environment has opened up a whole new range of service possibilities.

7.2 Data technologies

Data technologies fall broadly into circuit switched or packet data services—each with unique characteristics.

7.2.1 Circuit-switched connections

Circuit-switched data calls provide a temporary dedicated circuit between two points in a network. The circuit is temporary since it is only used for the duration of a call and dedicated because it is unshared by other users. Once the transmission is complete, the circuit is torn down and becomes available for other users.

Circuit-switched connections are efficient for voice traffic, which is long and interactive, and data transfer that exhibits the same characteristics. This would include fax transmissions or file transfers where data is being exchanged for a high proportion of the connect time.

Charges for circuit-switched calls generally follow a normal PSTN model—where the call is charged by duration of connect time regardless of the amount of data transmitted. Figure 7.1 shows a typical circuit-switched data connection.

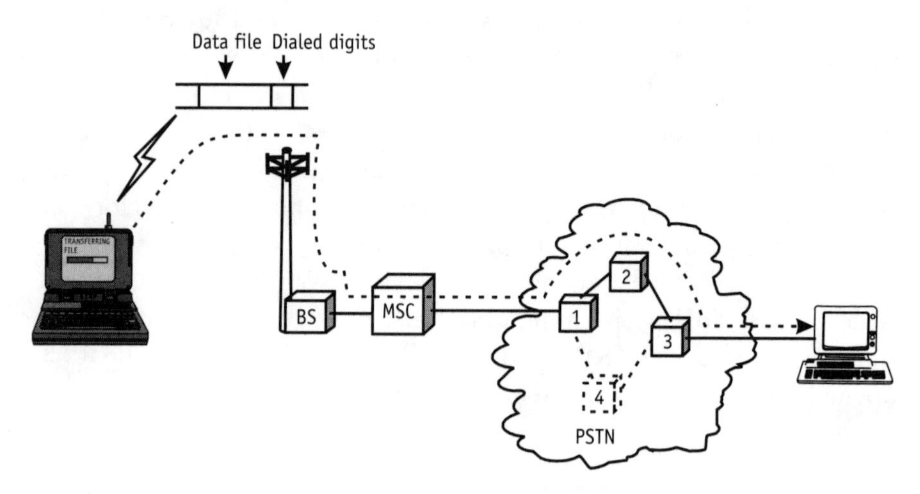

Figure 7.1 Circuit-switched data connection.

7.2.1.1 Wireless circuit-switched data

Circuit-switched data capability can use the analog cellular channels to transport data from 2,400–19,200 bps (greatly dependent on the radio conditions) across the cellular system, but a new approach must be used in an all-digital environment (for instance, at 1,900 MHz, where no analog channels exist). Owing to the voice coding schemes used on the digital traffic channels, it is very difficult to reliably use these digital channels for data or fax information. Two companion standards to IS-136, IS-130, and IS-135 define a new radio link protocol and command set that use the existing digital traffic channel structure and provide a reliable air interface protocol for the transmission of data and fax information.

7.2.2 Packet data connections

Packet-switched networks allow many simultaneous users to access multiple locations across the network. Each packet of data contains address or routing information. The packets can be delivered using many different routes across the network in order to find the quickest, least expensive, or most suitable route to the destination.

Each single transaction may result in many packets being created and transmitted across different routes in the network. The packets may be received out of order or damaged, and it is the responsibility of the underlying protocol to reorder the packets, request missing packets, and transfer correct data to the destination. Typical packet sizes range from 100 to 1,500 bytes of data.

Charging for packet transaction is based on the volume of data. This means that a device could be connected for an extended period of time and the user only charged for the actual data transmitted. Typical billing plans are flat-rate, with a user entitled to a certain number of kilobytes of data. When the user exceeds the entitlement, additional data is charged at a fixed per-kilobyte rate.

This makes packet data ideally suited to bursty transmission applications such as credit card authorization, dispatching, and database queries. Figure 7.2 shows a typical packet-switched data connection.

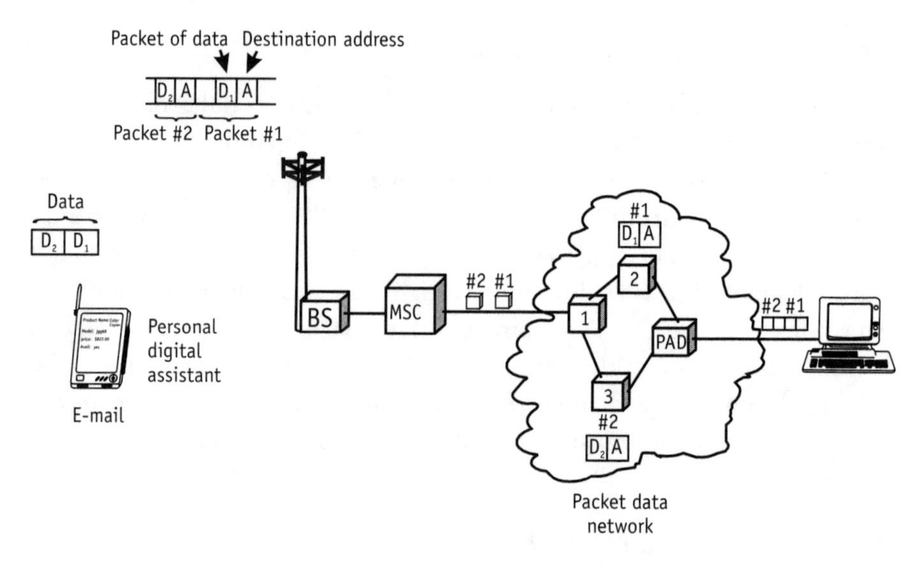

Figure 7.2 Packet-switched data connection.

7.3 Teleservices in IS-136

A special type of data technology exists in the IS-136 environment for the transport of teleservice information across the air interface. This technology, which is similar to a packet transmission, enables short packets of information to be sent to and from a wireless phone. Initial use for this technology was the CMT, which transports short text messages and displays them on the screen of a phone—much like a pager. For CMT, the message size can contain up to 239 characters. More advanced teleservices have been added to support OAP and UDP data transport.

A new node in the cellular system, called a message center, provides a store-and-forward capability for this information and can interface to a number of message sources (e.g., dial-up operator, TAP, TNPP, SMTP, SMPP, or the Internet) for more flexible information. The service can be broadcast (point-to-multipoint), mobile-originating, and mobile-terminating.

In addition to CMT, other teleservices can be provided on the same transport mechanism. For instance, OAP, OATS, and GUTS (used to support microbrowser technologies in a phone) can make use of this teleservice transport. Chapter 3 describes the delivery technology and acknowledged transport mechanism. Figure 7.3 shows an example of a teleservice used to deliver a text message to a phone.

7.4 Analog circuit-switched data

Analog circuit-switched data for cellular capability, which is sometimes provided with a cellular-capable PCMCIA modem card, enables the subscribers to use a cellular phone instead of a landline connection for dial-up data sessions and fax transfer.

Figure 7.4 shows a block diagram of this arrangement. In this case, the cellular phone is providing an analog audio channel across the air interface as a bearer for the data session. At the switch, the call is sent across the PSTN to the destination modem or fax machine. Figure 7.4 shows the typical analog cellular circuit-switched data connection.

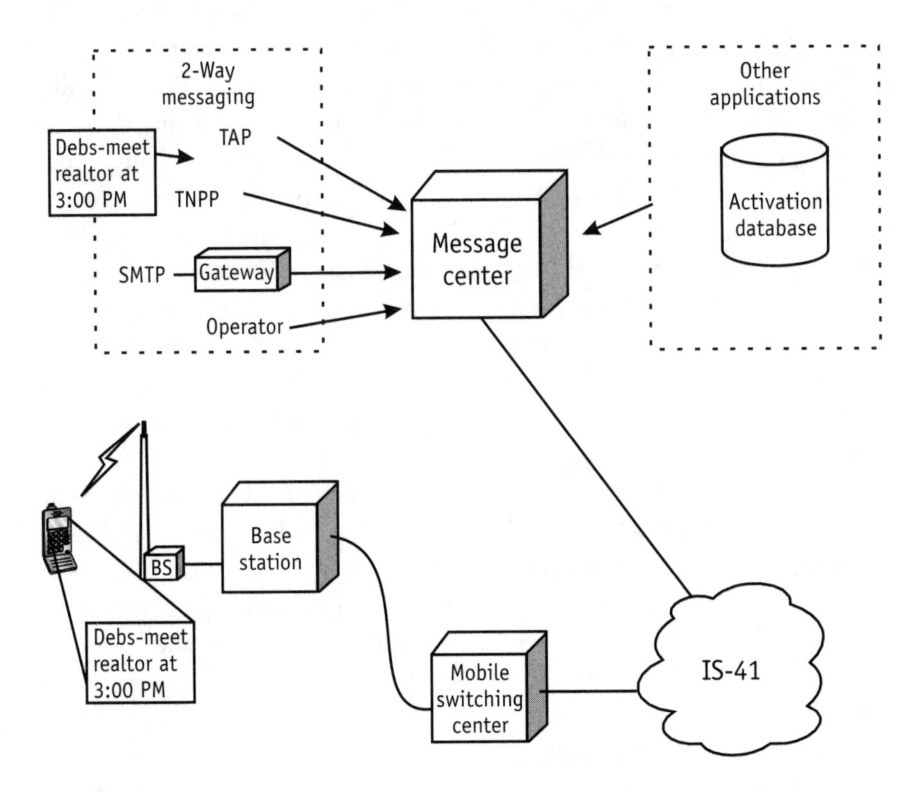

Figure 7.3 Teleservice.

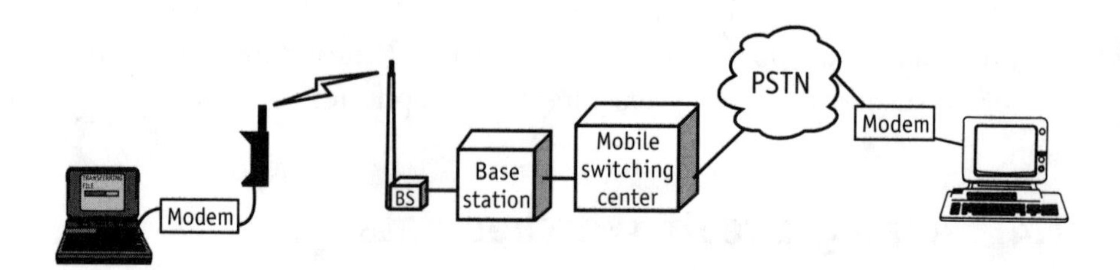

Figure 7.4 Analog cellular circuit-switched data.

7.4.1 Enhanced modem protocols (MNP10 and ETC, ETC2, TX-CEL)

To enhance the data transfer rate, several special modem protocols, which can adapt to the distortions that occur on the radio channels, have been developed. In some markets, modem pools installed at the cellular switch use protocols to enhance the reliability of the data path across the air interface. Two such protocols, *Microcom networking protocol class 10* (MNP10) and *enhanced throughput cellular* (ETC), increase the reliability of the data or fax connection by coping better with varying channel quality or the disruptions of the cellular environment (i.e., during hand-off).

These protocols are also available in landline modems for direct end-to-end error protection in cases in which modem pools have not been installed in the switch servicing the cellular call. MNP10 and ETC can also be used for more reliable landline-to-landline calls where channel quality degradations are likely (e.g., in rural areas or for international calling). Figure 7.5 shows how the MNP10 or ETC protocols can improve the connection for data transmission on an analog cellular system.

A drawback of the analog approach is that the data connection relies heavily on the quality of the underlying analog audio channel and radio artifacts; this would not normally be a problem for a voice call, but it can make the data connection somewhat fragile. In addition, the data cannot be easily transported over a digital channel in the same audio fashion due to the voice-coding techniques used by the digital voice coders as described in Section 7.5. This means that the analog data capability is limited to the 800-MHz cellular coverage areas and is not available in 1,900-MHz systems in which no analog voice channels are present.

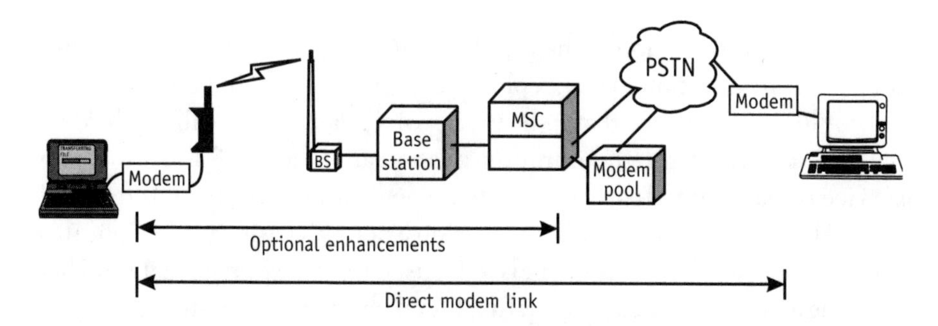

Figure 7.5 Enhanced analog cellular circuit-switched data.

7.5 Digital circuit-switched data and fax services

As discussed in Section 7.4, digital traffic channels are not ideally suited for transferring data or fax information because of the channel and voice coding that is performed in the VSELP or ACELP processes. These voice coders are designed to transmit symbols and codes that represent a tremendous amount of compressed voice data—including pitch and volume wave shape. In addition, the most perceptually significant information is further coded to avoid unwanted errors from affecting the audio output.

In order to use the digital traffic channels for data and fax communication, a new approach was adopted. This split the functionality of a regular modem into two parts: The digital interface to the PC was kept at the mobile side, and the analog PSTN data pump functions were moved to the switch. The two halves were joined by an error correcting radio link protocol (RLP1) that was designed to cope well with the harsh radio environment.

Instead of attempting to transmit audio information across the air interface, RLP1 sends *protocol data units* (PDUs) or error-protected packets from the phone to the system. The switch decodes the PDUs and transfers the data to the data pumps where it is sent over the PSTN in a normal audio manner. The reverse operation takes place in the downlink direction.

This data transport scheme was designed to use the existing IS-136 DCCH call control features and ancillary processes but to replace the voice codec information in the regular TDMA digital traffic channel with the RLP1 information. Thus, a phone camping on a digital control channel could receive a page that indicated a voice, data, or fax call and be designated to a DTC for the duration of the call. The benefit of this data service is that channels do not have to be dedicated, since the same digital channel can be reused for voice or data on a per-call basis. In addition, this mechanism allows digital channels to be used for transporting data and fax with the associated capacity benefits, hierarchical cell structures, and private system features of an integrated IS-136 approach.

RLP1, specified in IS-130, is accompanied by IS-135, which is the AT command set and specification for user data transport. Together, these provide the digital circuit-switched data and fax capability for IS-136.

The digital circuit-switched data and fax service provides a reliable transport mechanism for user data and fax information in an IS-136 environment. The raw data throughput is 9,600 bps on a full-rate channel (that is, on the same bandwidth used by a voice call on a full-rate DTC). Compression is provided on the air interface and can improve the data rate up to four times to 38,400 bps, depending on the radio conditions and the compressibility of the data. Mobile-originating and mobile-terminating services are specified for both the data and fax services. Figure 7.6 shows the transfer of data using a full-rate channel. This allows reliable data transfer at 9.6 Kbps.

More than one full-rate traffic channel can be concatenated to provide enhanced data transports called double- or triple-rate channels. These use two or three full-rate channels together and increase the basic bit rate to 19,200 and 28,800 bps, respectively. Again, compression can further enhance the throughput for the subscriber.

The fax service uses the class 2.0 fax standard to provide group-3 fax capability. This gives a fax throughput of approximately two pages per minute with error-corrected performance for clean delivery.

Data encryption is provided to protect the privacy of the data sent across the air. This process uses information from the authentication process to encrypt the transmitted data.

The phone is designed to interface to a PC or other device directly using a standard serial port (V.24) much like an external modem does today with no additional modem being required. Thus, the phone can be designed to achieve a close similarity to other modem devices from

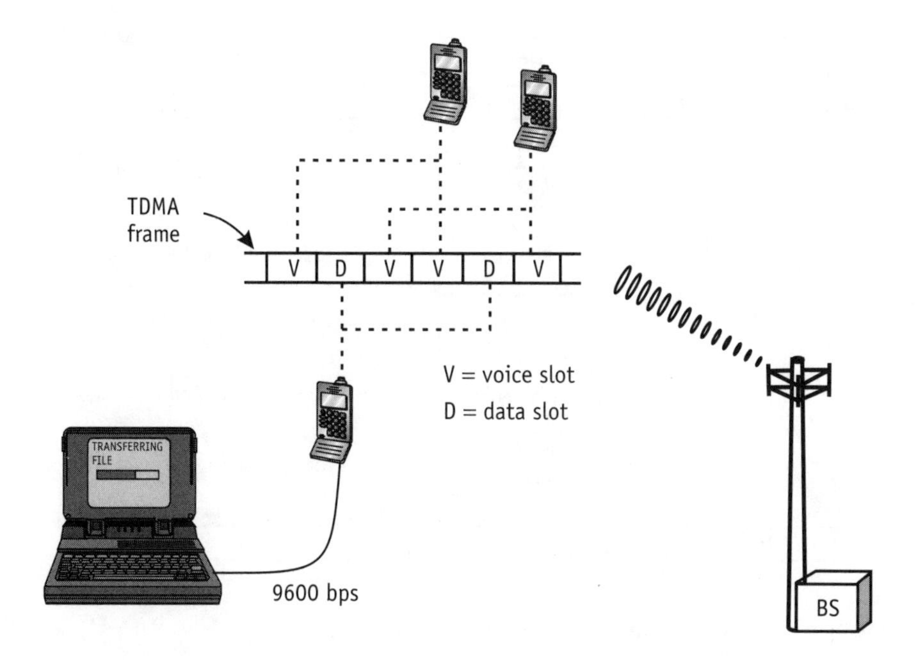

Figure 7.6 Full-rate data.

the PC point of view, as well as to support the plug-and-play® architecture of PC operating systems. This greatly simplifies the PC-to-phone connection for the data user.

7.5.1 The radio link protocol (IS-130)

The radio link protocol of IS-130 provides a reliable data channel for the transport of IS-135 information. The protocol consists of seven main functions described here and shown in Figure 7.7.

Octets of information are received from the higher layers and compressed into a smaller number of codewords. The compression standard is the widely used V.42bis protocol. Up to four times compression can be achieved with V.42bis, depending on a number of factors, including device memory resources, radio conditions, and the compressibility of the data (that is, if the data has already been compressed or otherwise manipulated).

Compressed codewords are then blocked together and sent to the transmission controller. This encrypts the data if requested, adds control

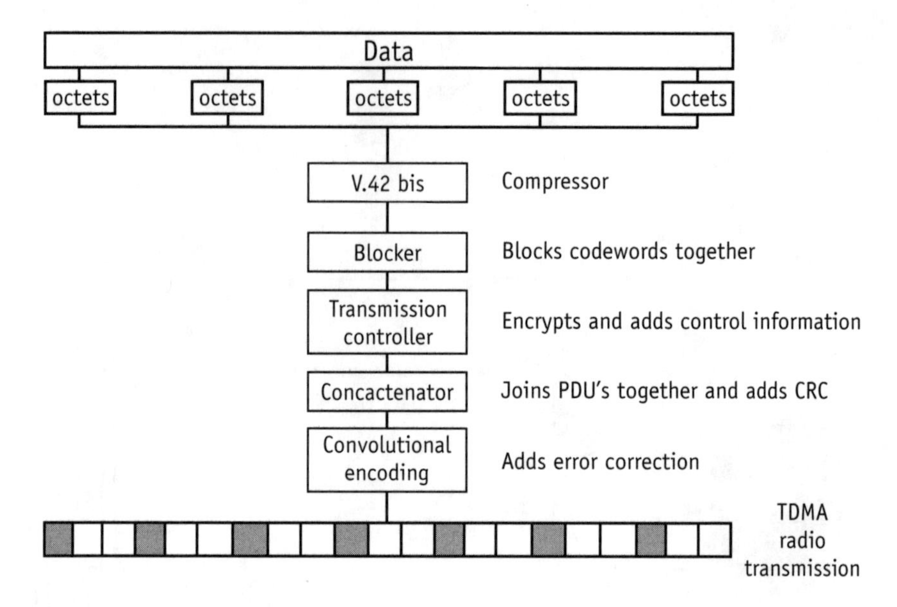

Figure 7.7 RLP1: The radio link protocol.

information, and outputs a PDU. The encryption is performed using information generated by the authentication process. The control information is used by the receive transmission controller on the other side of the radio link to request retransmission of data that was not received correctly.

PDUs are again joined together in the concatenator, a frame check sequence consisting of a CRC field is calculated, and the resulting block of data is convolutionally encoded to add further error protection. Subsequently, these RLP1 encoded frames are sent to the physical layer for transmission across the radio interface.

On the receive side, information from the physical layer is convolutionally decoded, the frame check sequence is calculated, and errored frames are removed. The separator splits up the PDUs, which are then passed to the transmission controller. Missing frames are requested from the transmitting side, and the correct frames are decrypted, deblocked, and expanded before the octets reach the higher layers.

In this way, a reliable data path for the transport of IS-135 and user information is established and maintained between the phone and the system.

7.5.2 The data specification (IS-135)

The IS-135 data specification defines how to set up, supervise, and release digital data and fax calls using the logical links, acknowledged data transport, compression, encryption, and flow control provided by RLP1. The mapping of the AT command set is also specified for basic asynchronous data and fax calls. The AT command set also includes some cellular-specific commands such as signal strength and battery status so that new PC applications can reflect phone conditions to the user.

IS-135 also specifies the connection between the PC or attached device and phone. This is defined as a serial connection sending characters of information across a V.24 serial interface.

7.5.3 Components of a digital circuit-switched data system

Figure 7.8 illustrates the main components of a system supporting digital circuit-switched data. The main addition is the *interworking function* (IWF) installed at the switch. It is this unit that is the other half of the conceptual split modem and that terminates the radio link protocol of RLP1. This unit also contains the data pumps for PSTN data and fax operation.

The remaining parts of the system are the same as a voice system with the switch, HLR, and base stations performing the same functions. In this case, the radios themselves will pass RLP1 information back to the IWF rather than VSELP or ACELP voice information back to a vocoder.

Mobile-originated calls start with the AT command and dial string being sent from the PC and phone to the system for processing. This uses

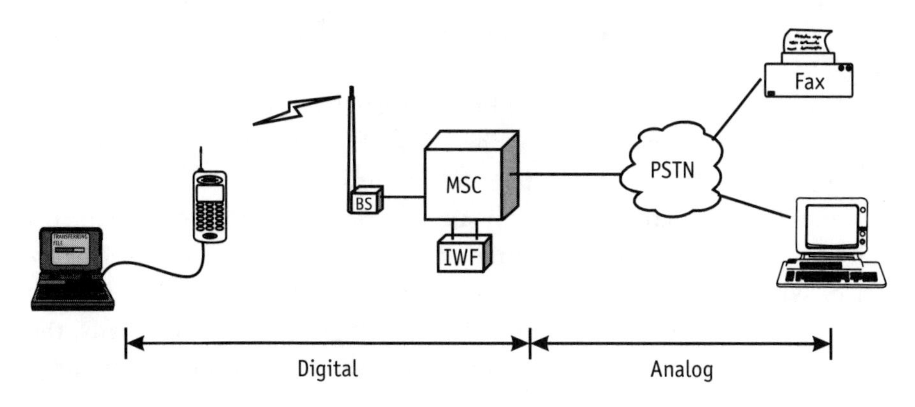

Figure 7.8 Digital circuit-switched data connection.

the control channel for initial origination signaling followed by the designation of a DTC for the notification and negotiation of IS-130 RLP1 and IS-135 transport information. This may include encryption and compression capabilities as well as autobauding, parity, start, and stop bits.

The cellular system, knowing that this is a data or fax call, passes this information through to the IWF instead of switching in a voice codec for this call. Data transfer can then take place once the lower link across the air interface has been established. The phone and IWF, therefore, work together to maintain a reliable data connection for the duration of the call.

It should be noted that all IS-136 features and services described in Chapter 10 are available whether the phone is being used for voice or data; that is, the extended battery life, private system capabilities, SMS, and other teleservices, and the HCS for in-building systems and traffic management are fully integrated in this new device.

7.5.4 Mobile devices and user model

The digital circuit-switched data and fax service is intended for operation with an attached piece of equipment—either a PC or PDA—that runs the application requiring wireless connectivity. The data standards specify connectivity using the V.24 protocol between the PC and phone, with the latter interface looking much like a standard AT modem for the PC point of view.

This configuration gives users the look and feel of their desktop environment with the advantage of wireless access. For example, on a laptop, this allows the same applications to be used at the office, in the hotel room, or at the airport, depending on the availability of Ethernet, PSTN, or IS-136.

In addition to the traditional PC, *personal digital assistants* (PDAs) and other hand-held devices such as the Win CE platform or Palm Pilot are gaining popularity with the advent of processors and operating systems designed for these compact platforms. Since IS-135 supports the standard AT commands, most basic functions should be possible on existing hardware, with the phone emulating a standard modem. In addition, with the wider adoption of plug-and-play technology in newer platforms, the set-up and use of these phones and devices should become rudimentary for more advanced operations. Figure 7.9 shows a PDA with communications capability.

7.5.5 Direct IP connectivity

The basic IS-135 service is designed for PSTN data and fax operation. An all-digital end-to-end solution can also be achieved using the IS-135 data service in conjunction with a direct IP connection from the switch to an *Internet service provider* (ISP) or corporate intranet. Instead of the data leaving the switch and being routed toward the PSTN, the data is sent directly over a frame relay or IP network to the far-end destination.

This network configuration has a number of advantages. The connection time is greatly reduced since no PSTN modem training is needed to establish the network connection. The scalability of the network is also improved since many circuits can be concentrated over the IP network

Figure 7.9 PDA with communications capability. (*Photo courtesy of:* Casio.)

without using a PSTN circuit for each data call. Thus, trunking to an ISP, for example, can be much more cost-effective.

The standard Internet protocol suite can be used in this configuration, as well as the established authentication and security measures that are in place today.

7.6 Packet data in IS-136

Packet data enables many users to access many locations across a network. Packets, which are sent from source to destination using the fastest route available, share a resource that is normally occupied by a single user. It is desirable to extend this concept into the radio environment so that many users share the same radio channel for packet data transactions rather than one user per traffic channel as in the circuit-switched case.

The *cellular digital packet data* (CDPD) technology uses this approach to provide a cellular packet data service on the 30-kHz cellular radio channel normally used for analog voice channels. CDPD uses a different modulation scheme (GMSK) than AMPS or IS-136, but many AMPS/CDPD devices are available today. CDPD can easily be overlaid into an existing 800-MHz AMPS, IS-54B, or IS-136 network.

For the pure IS-136 environment, a coding scheme and architecture is being developed to map the packet protocols into a digital channel environment. In this way, a fully integrated packet data solution can be realized within the framework of the existing IS-136 infrastructure. One major advantage to having the packet data solution integrated into IS-136 is that phones and data terminals are greatly simplified, and the chances of losing an incoming call when a phone is selecting between the two systems is eliminated. In addition, integration into the IS-136 environment continues to provide the key IS-136 benefits of the sleep mode, private system, and hierarchical cell features.

7.7 Data applications

Existing applications that would use a wireless data transport tend to fall into the circuit and packet arenas, but new applications are being developed that can use a mix of both circuit, packet, and teleservices, depending on application requirements.

Data applications that are ideally suited to circuit data are those in which data is being transferred for a high proportion of the connection time, including fax, e-mail bulk retrieval, and file transfer. Connectionless applications such as credit card verification are more suited to a packet approach owing to the bursty nature of the transmission. Online e-mail editing is also better served by packet as it is also bursty in nature. Teleservices provide a straightforward method of providing a packet-like data delivery for low-volume, sporadic data delivery for e-mail headers and alphanumeric paging functions.

7.8 Browsing the Web

With the rapid growth of the Internet and associated technologies, methods to leverage these technologies in the wireless environment have been pursued. Eventually, the flourishing IP-based applications will move to the wireless arena, making packet-oriented wireless data increasingly more important. The browsing paradigm of HTML, in which pages of information reside in a network and are accessed by a client application, is one example. This can be created in a wireless environment with network-based applications and clients residing in the wireless phone or data device to provide an extremely flexible architecture for future services.

The *hand-held device markup language* (HDML), created by Unwired Planet Inc., is an offshoot of HTML; with the associated *handheld device transport protocol* (HDTP). HDML enables "cards" or pages of information to be exchanged between a network application and the phone as shown in Figure 7.10. HDML is optimized for low-bandwidth operation using limited screen size and limited user input keys. The cards can consist of information, choices, or input capabilities, with the ability to dynamically define the soft-key function of the phone and return the key selections to the network application.

The HDML model can be used to devise extremely flexible applications that reside in the network and are accessed by browser-capable wireless devices. These applications may provide news, weather, stock

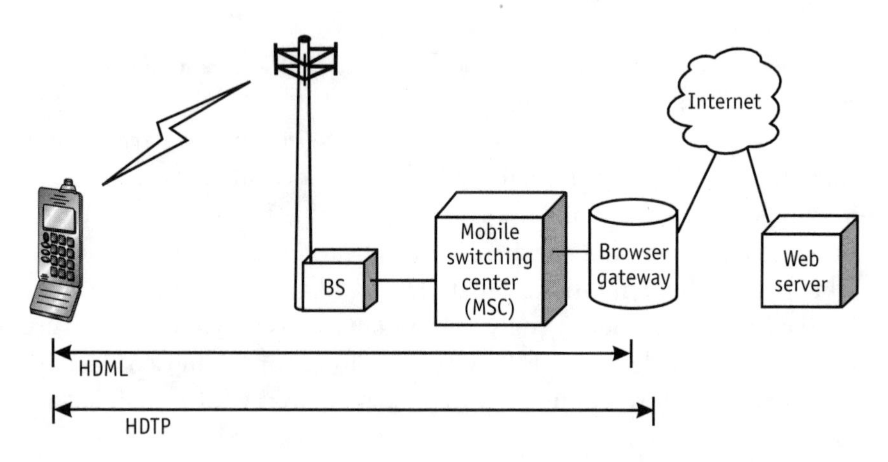

Figure 7.10 Connecting to the Web using wireless phone.

quotes, or e-mail capability. In addition, they are fully programmable so that end users or corporations can tailor the applications to suit their individual needs.

7.9 The thin client architecture

The thin client architecture, which is an example of the browsing paradigm, enables a wireless device to use available networks for exchanging browser information. The current architecture can be served by the teleservice transport mechanism, a digital circuit-switched transport, and CDPD. Future IS-136 packet data services will also provide a bearer service for this architecture. Figure 7.11 shows the thin client architecture.

7.9.1 Service overview

The exchange of information between the system and the phone can fall into one of three categories listed as follows:

- Alerts and notifications: The system alerts the user to pending e-mail, etc.
- Two-way value-added services: The user can interact with the system for two-way messaging or other short, low-volume transactions.

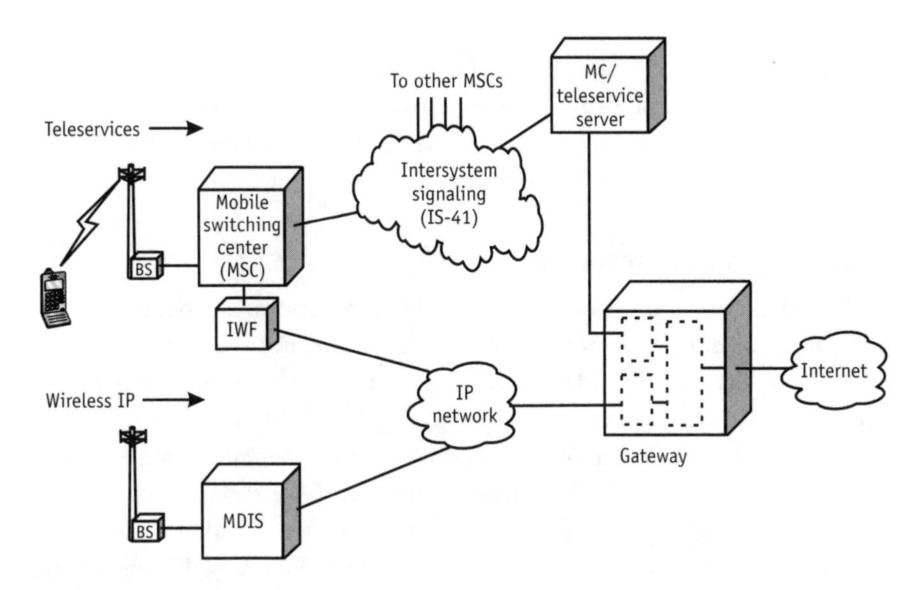

Figure 7.11 Thin client architecture.

- Wireless IP for full browsing: Users can download their e-mail or browse the Web where data volume is uncontrolled.

7.9.1.1 Alerts and notifications

The alerts and notifications (indications that information is pending) are transported over the IS-136 GUTS teleservice, which provides a mechanism to deliver UDP information over a DCCH. The phone can store this information for immediate action by the user or for action at a later time.

7.9.1.2 Two-way value-added services

The user is able to undertake limited operations of value-added services applications that use this teleservice without having to invoke the wireless IP mode. This would be used for small, low-bandwidth, controlled applications. Possibilities include the following.

- Mobile-originated and mobile-terminated messaging;
- Call forwarding and other network or user profile features;
- Limited, controlled applications (such as news headlines and weather) that would not over burden the DCCH.

7.9.1.3 Wireless IP for browsing the Web

The wireless IP data transfer mechanism can use whatever network resource is available; this could be IS-135 digital circuit-switched data with IP direct connection, CDPD, or IS-136 packet service.

7.9.2 Transport protocols

As described previously, the HDML/HDTP protocols can be used in the thin client architecture. In addition, the emerging *wireless application protocol* (WAP) can be used to provide the client browser, network transport, and gateway/server functionality. The WAP is a client-server architecture being developed by a consortium of terminal vendors and system operators. WAP takes into account different classes of devices with different display/MMI capabilities and focuses on value-added teleservices integrated with browser functionality. Figure 7.13 shows the protocol stack for a browser in an IS-136 environment.

7.9.3 Teleservice transport

To transfer data while the phone is in idle mode (camping), GUTS is used. GUTS provides a transparent pipe for UDP information between the phone and system and uses the same R-data transport mechanism as the other IS-136 teleservices.

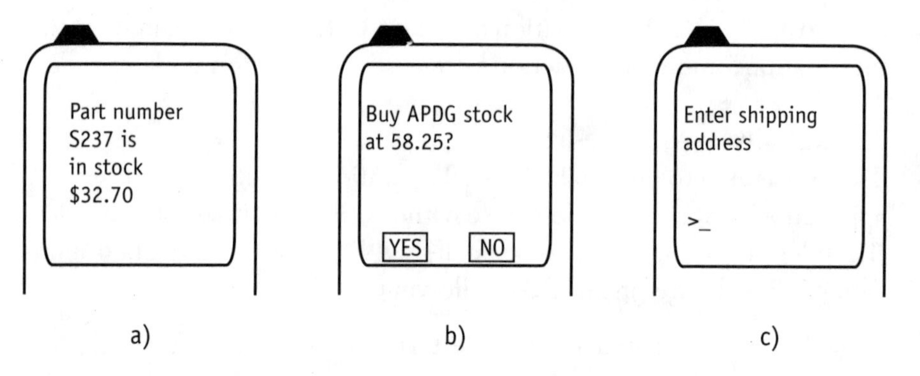

Figure 7.12 Browsing the Web using wireless phone: (a) display, (b) choice, and (c) input.

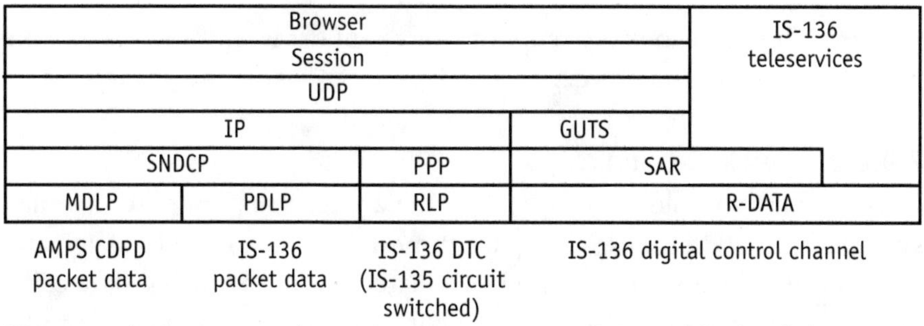

Browser
Session
UDP
IP | GUTS
SNDCP | PPP | SAR
MDLP | PDLP | RLP | R-DATA
IS-136 teleservices

AMPS CDPD packet data | IS-136 packet data | IS-136 DTC (IS-135 circuit switched) | IS-136 digital control channel

UDP = user datagram protocol
IP = internet protocol
GUTS = general udp teleservice (IS-136)
SAR = segmentation and reassembly (IS-136)
SNDCP = subnetwork dependent convergence protocol
MDLP = mobile data link protocol
PDLP = packet data link protocol
PPP = point-to-point protocol
RLP = radio link protocol
R-DATA = relay data

Figure 7.13 Browser protocol stack.

7.9.4 Circuit-switched data transport

When wireless IP calls for a circuit connection, the digital circuit-switched data capability of IS-130 and IS-135 can be employed with the direct IP connection rather than PSTN connectivity to the network complex. This provides a fast set-up time and excellent trunking efficiencies though the network for the IP-based traffic.

7.9.5 Packet transport

Both CDPD or future IS-136 packet services can be employed for packet transport depending on network infrastructure and suitably equipped phones.

7.9.6 Components for the thin client

7.9.6.1 Wireless device

The wireless device in the thin client architecture can be a wireless phone, a PDA with a browser and wireless modem, or a similar device. The phone is required to support IS-136, two-way transport teleservices, and either IS-135, CDPD, or IS-136 packet for wireless IP operation. In addition, the phone needs to support the client browser for managing the information from the server.

7.9.6.2 Network complex

The network complex needs to have the ability to operate in a circuit-switched IP, CDPD, or IS-136 packet mode and may need to assign and maintain IP addresses for mobile-originated circuit-switched data calls. The network complex also needs to contain a proxy function and to be able to translate Internet-sourced IP addresses into MINs and to interact with the message center to deliver alerts and notifications to phones using GUTS from the GUTS-capable message center/teleservice server.

7.9.6.3 GUTS-capable message center/teleservice server

The GUTS-capable message center/teleservice server is able to deliver GUTS messages to browser-capable phones using MIN addressing. The message center is also able to receive mobile-originated GUTS messages and to route them back to the complex.

7.10 Data devices

In addition to mobile phones, other wireless devices are capable of providing wireless data functionality.

7.10.1 Dedicated data radios

Dedicated data radios are modules that are designed to be embedded into larger devices to create a wireless data product. For example, a PDA might be integrated with a built-in data radio to create a wireless PDA.

7.10.2 Data adapters

A data adapter is a cable or interface box that adapts an existing phone into a data-capable phone. The data adapter provides a serial connection to a PC or PDA and uses analog circuit-switched data as the radio bearer.

7.10.3 Wireless modems

A wireless modem is an all-inclusive device that converts serial data into digital radio signals. Wireless modems are available as standalone devices or as type-2 or type-3 PCMCIA cards with external antennas.

7.11 Future IS-136 data activities

With data capabilities demanding higher bandwidths, faster connections, and more integration with voice services, several enhancements are being pursued within the mobile phone industry standard organizations.

7.11.1 Multislot operation

Multislot operation enables TDMA time slots to be concatenated to give a higher throughput on the air interface as shown in Figure 7.14. This means that a raw bearer of 19,200 or 28,800 bps could be provided on double- or triple-rate TDMA channels, respectively.

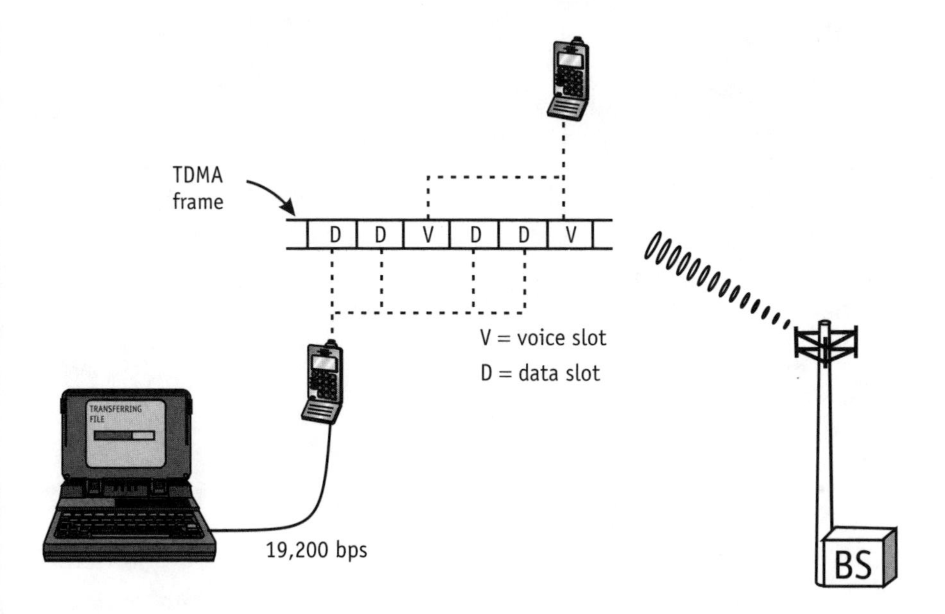

Figure 7.14 Double-rate data.

7.11.2 New modulation schemes

Different modulation schemes can improve the number of bits represented by each phase shift to increase the overall bandwidth. Various schemes are under consideration to increase the information throughput but maintain the same TDMA structure of the radio interface.

7.11.3 Packet data transport in the IS-136 environment

Various schemes are in development to provide an integrated packet data solution for TDMA. In the short term, the TDMA community is targeting 45-Kbps packet data rates; in the longer term, the 384-Kbps target has been clearly identified.

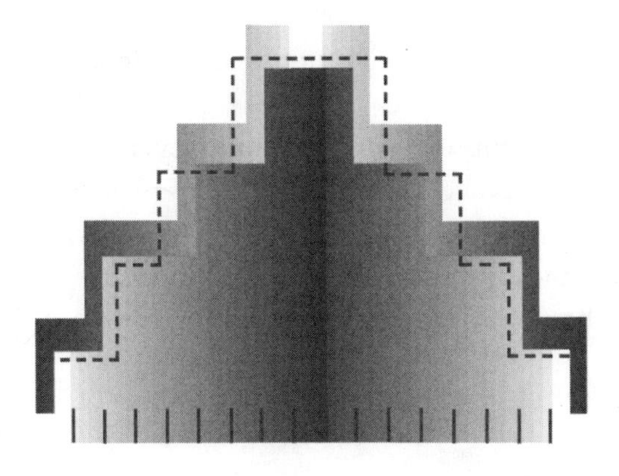

8

Dual-Mode, Dual-Band Phones

8.1 Introduction

This chapter covers dual-mode/dual-band phones, which can use either analog or digital voice channels and operate at both the 800-MHz and 1,900-MHz frequency bands. The phones, which are now available, give the maximum coverage to the subscriber while operating in the most efficient mode—depending on network availability.

8.2 Analog and digital phones

Analog cellular phones designed to the AMPS specification have been prevalent around the world, with over 70 million handsets in use today. These phones use an ACC, which is created using FSK on a 30-kHz channel to provide the air interface signaling between the phone and system.

The voice path is provided by frequency modulating the audio signal on 30-kHz channels.

These phones provide a basic incoming and outgoing service (see Figure 8.1) with little opportunity to take advantage of the newer network features, such as SMS or sleep mode. They can, however, provide enhancements as component technology and memory constraints relax and enable a more fully featured handset with improved characteristics.

8.3 Dual mode

IS-54B introduced *digital traffic channels* (DTCs), which enabled the voice channel to be shared by up to three users using full-rate channels, or six users using half-rate channels. IS-54B used *time division multiple access* (TDMA) to slice the channel into time slots that can be assigned to different users.

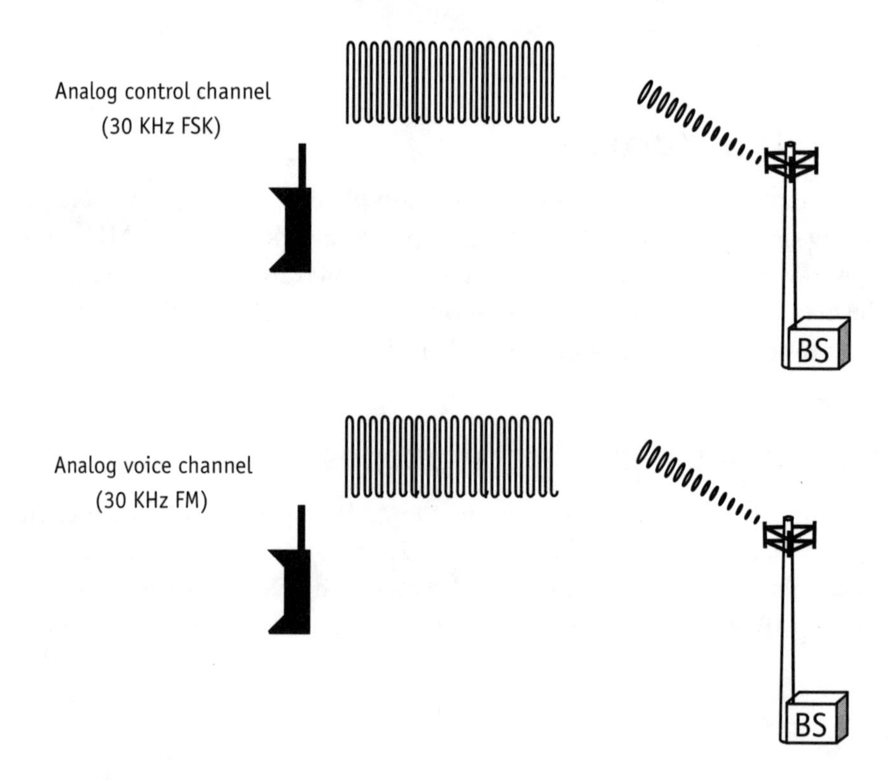

Figure 8.1 AMPS service.

Dual mode is the term used for phones that can operate using either analog voice channels or DTCs (see Figure 8.2). This allows the subscriber to take advantage of the digital system or analog system, depending on system capability, and maximizes the available coverage area for the user. This is especially useful when systems are being built out or when the digital subscriber roams into an analog area.

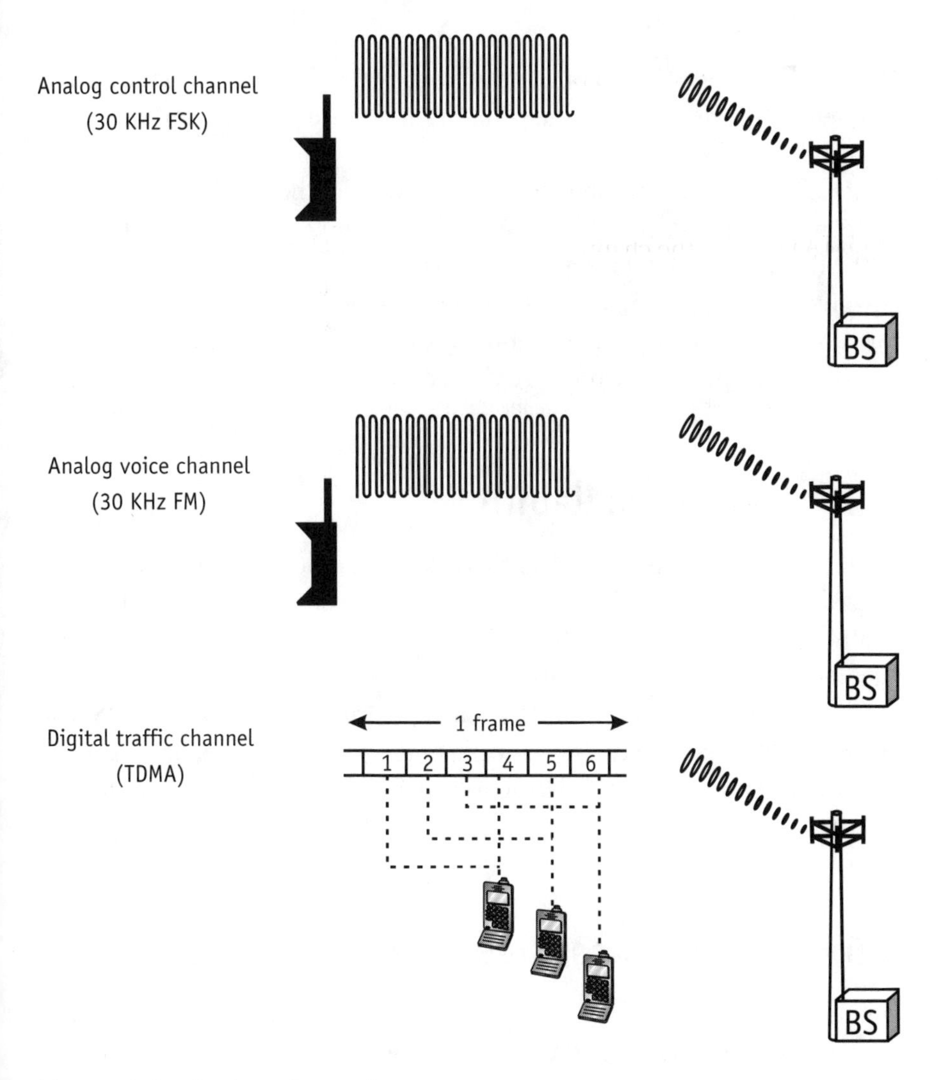

Figure 8.2 Dual-mode service.

IS-136 adds a further channel type to the spectrum by redefining the first time slot pair in any given radio in a sector to be the DCCH, which provides enhanced services and new capabilities to the subscriber. See Figure 8.3. Hence, dual-mode phones can also take advantage of the enhanced services offered by the DCCH.

8.4 Dual band

IS-136 is also specified to operate at both the 800-MHz and 1,900-MHz frequency bands with exactly the same features and services. See Figure 8.4.

Dual band refers to phones that can utilize either 800 or 1,900 MHz for communication (see Figure 8.5). Hence, phones can use either the 800-MHz or 1,900-MHz band depending on the frequency band used in any given geographic area. This provides a seamless PCS service offering for subscribers as they roam in and out of different service areas, regardless of the frequency used to provide service.

8.5 Feature availability

As discussed previously, IS-136 provides a seamless feature and service coverage across 800 MHz and 1,900 MHz. This means that all features are unaffected by the frequency of operation where IS-136 is present at both frequencies referring to Figure 8.6. Analog-only systems provide basic voice communications. IS-54B adds the availability of DTCs. IS-136 enhances this service with the addition of longer battery life, SMS, private system functions, and the seamless dual-band operation described in Chapter 10.

8.6 Dual-band consideration

Two aspects of the phones' operation, reselection and hand-off, deserve a more detailed description when considering a dual-band environment. See Figure 8.7. Reselection is the process by which the phone

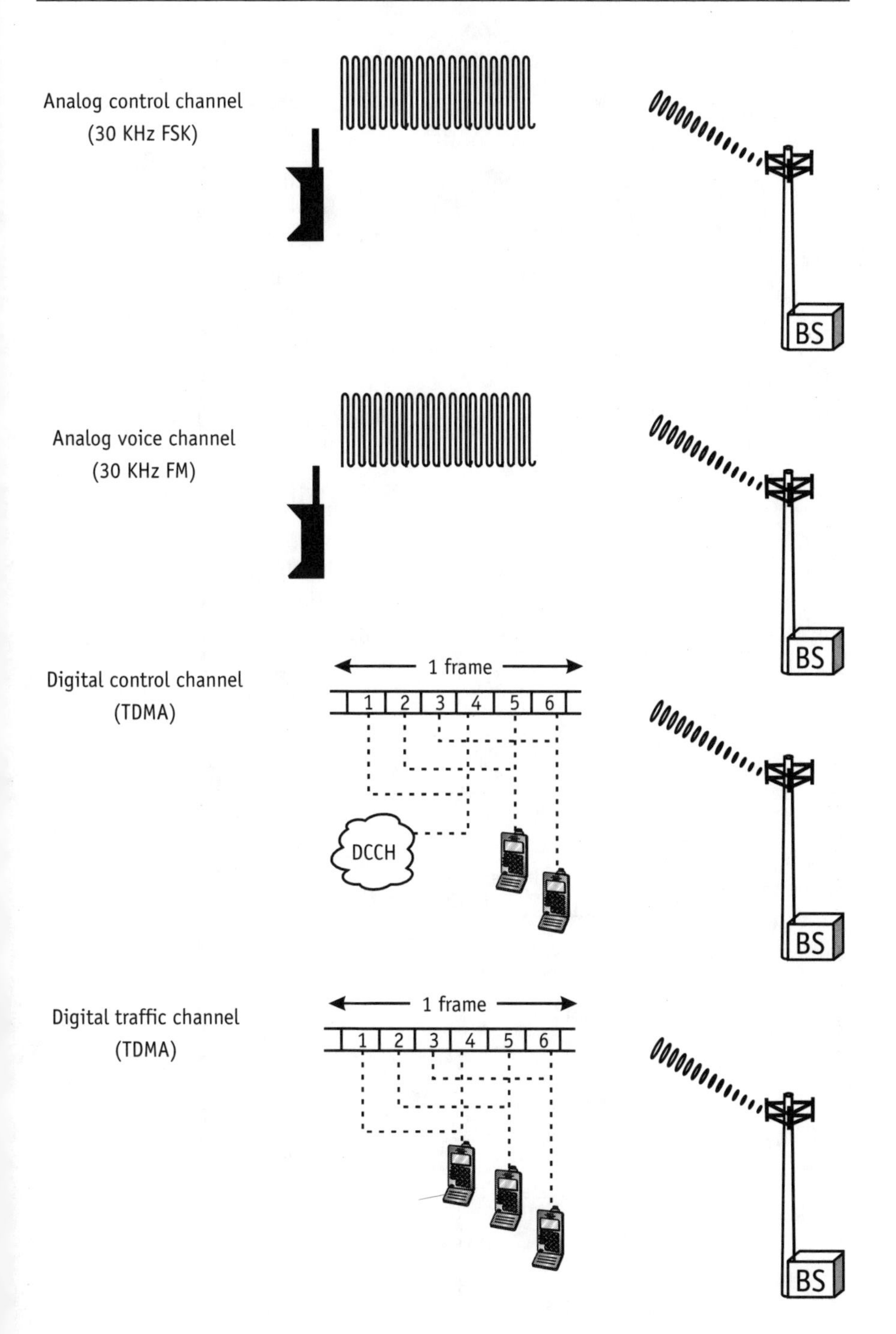

Figure 8.3 IS-136 dual-mode service.

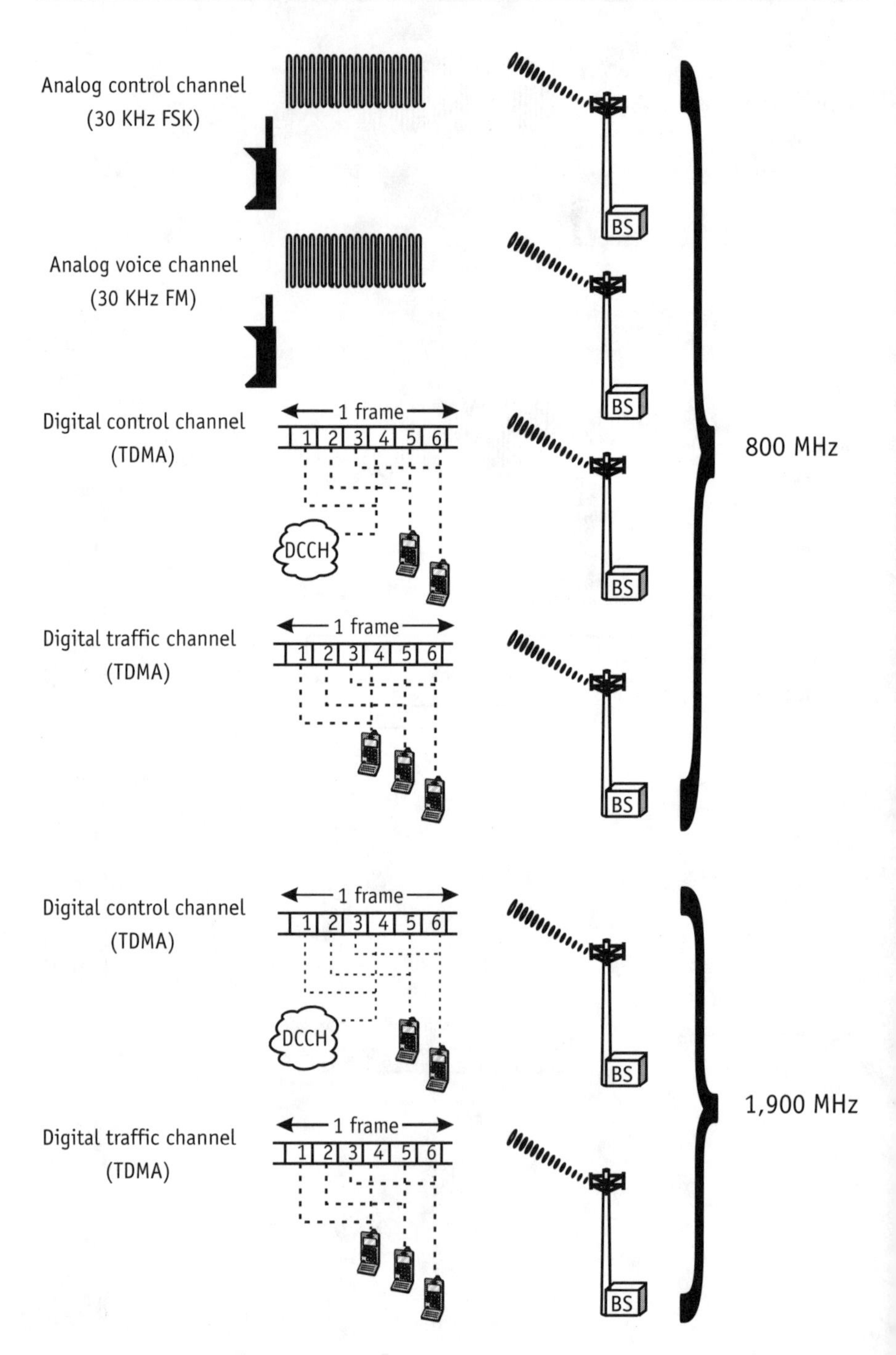

Figure 8.4 Dual-mode and dual-band service.

Figure 8.5 Dual-band phone. (*Photo courtesy of:* Ericsson.)

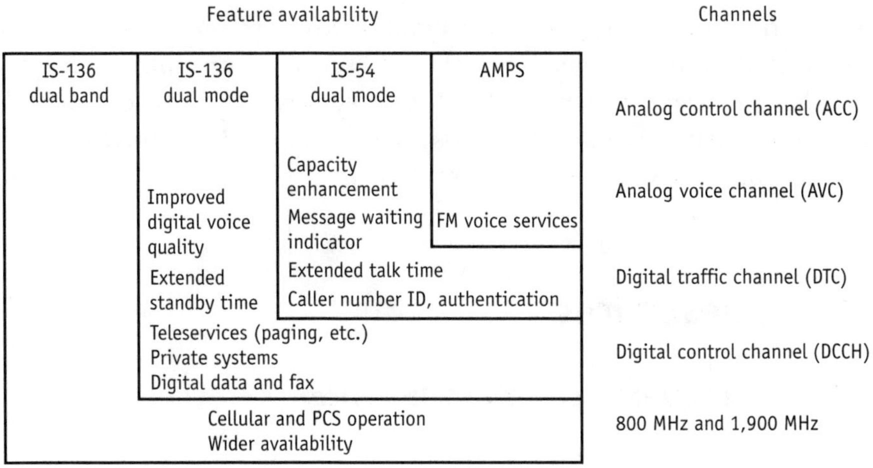

Feature availability

IS-136 dual band

IS-136 dual mode

IS-54 dual mode

AMPS

FM voice services

Capacity enhancement
Message waiting indicator
Extended talk time
Caller number ID, authentication

Improved digital voice quality
Extended standby time
Teleservices (paging, etc.)
Private systems
Digital data and fax

Cellular and PCS operation
Wider availability

Channels

Analog control channel (ACC)

Analog voice channel (AVC)

Digital traffic channel (DTC)

Digital control channel (DCCH)

800 MHz and 1,900 MHz

Figure 8.6 Feature matrix.

continuously measures, evaluates, and chooses the most suitable control channels (both DCCHs and ACCs) from the neighboring cell sites.

In the hand-off process, a phone is ordered to a different traffic channel during a call. Both of these processes can be performed between the frequency bands. That is, an 800-MHz IS-136 system can have a 1,900-MHz overlay system or neighbor cells and vice versa. Reselection

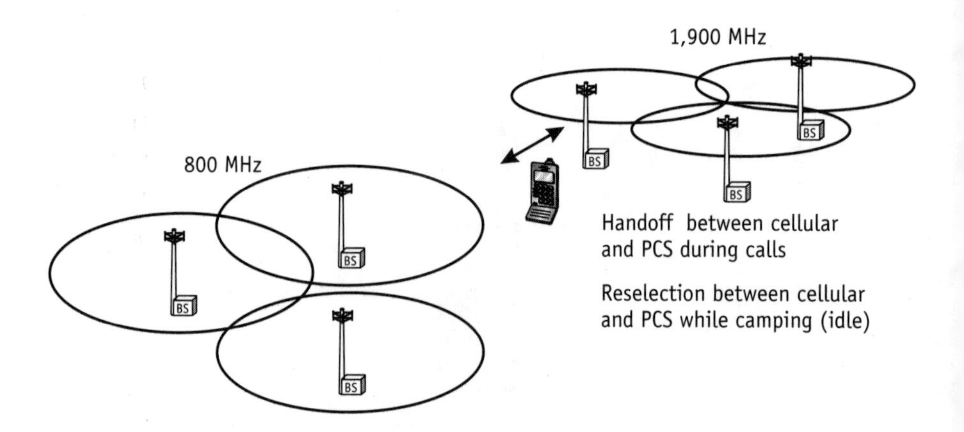

Figure 8.7 Dual-band considerations.

and hand-off can occur in both directions, allowing a seamless system to be created across frequencies. This is especially valuable where an 800-MHz system borders a 1,900-MHz system. The reselection and hyperband hand-off capability allows carriers to provide a seamless roaming environment with neighboring systems.

8.7 Teleservices

8.7.1 Over-the-air activation teleservice

OAA simplifies the phone activation process—for both the subscriber and service provider—by delivering NAM information and updates to the mobile station over the air without the need for "manual" intervention or the programming of the phone by the subscriber. See Figure 8.8.

The NAM information, which is downloaded to the phone during the activation process, includes the phone number, system identities, and other NAM contents. In this way, a flexible download capability for delivering service-provider or manufacturer-specific data, as well as in-the-field updates, is provided.

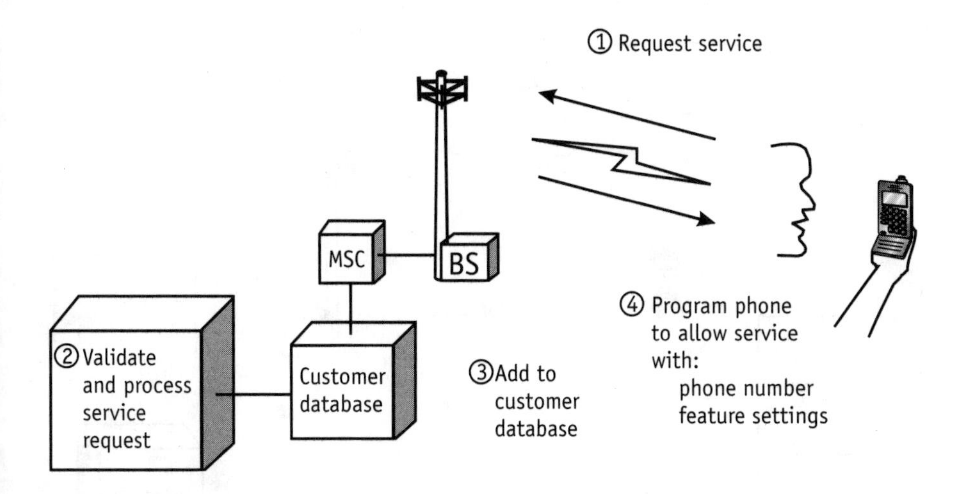

Figure 8.8 OATS.

8.8 Over-the-air programming and intelligent roaming

The OAP teleservice (see Figure 8.9) is used in conjunction with a mobile scanning mechanism called intelligent roaming. In a traditional cellular environment in the 800-MHz frequency band, there were originally two cellular providers—the nonwireline and wireline service provider operating the A and B band, respectively. With the popularity of PCS services at 1,900 MHz, there are now many service providers in any given geographical area, some with and some without compatible air interface technologies or roaming abilities.

Intelligent roaming is a process that allows the phone to seek out its own service provider from all the other sources of RF energy in the two 800-MHz and six 1,900-MHz frequency bands. If the subscriber's own service provider is not available, the phone must know which other channel's band will provide service to the subscriber. Alternatively, the phone must be intelligent enough to continue searching if a control channel not belonging to the subscriber's service provider is found. However, the subscriber's service provider may be in another PCS band in the same geographical area. A prime consideration in this process is that the

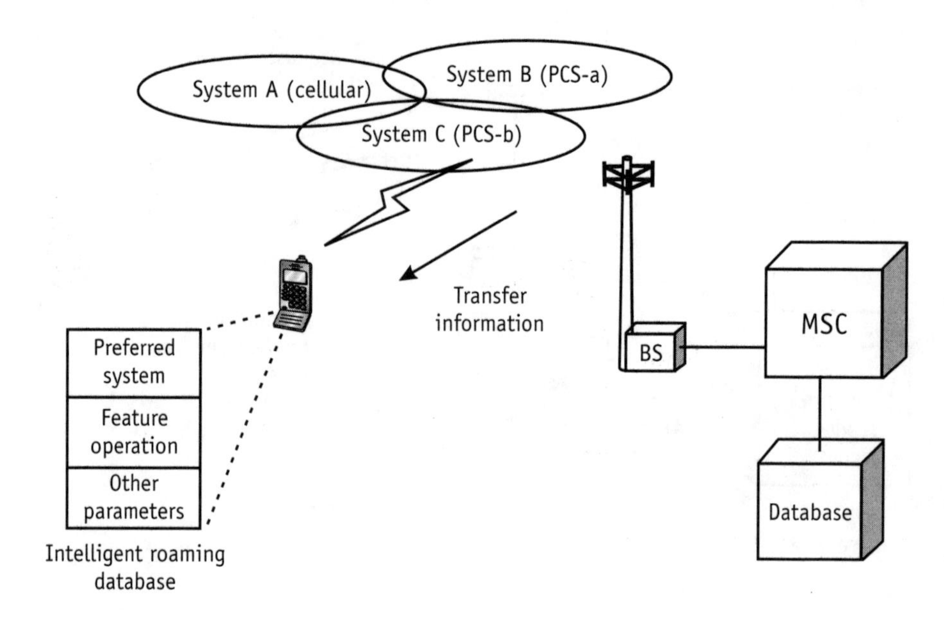

Figure 8.9 OAP teleservice.

channel scan is completed quickly so that the subscriber is capable of making or receiving calls as soon as possible after the power is on.

OAP is the downloading of the *intelligent roaming database* (IRDB), which is a list of SIDs, SOCs, the frequency band (800 or 1,900 MHz), and preferences. This list, which is maintained by the service provider, can be downloaded to the phone at service activation, or when the information requires updating. The phone consults this list when it first powers on to determine the most appropriate scanning strategy. This controls whether the phone scans the A or B 800-MHz band to look for service or whether it starts to search the 1,900-MHz bands. Not all bands need be searched; the IRDB lists which bands will be searched in what order.

Accordingly, subscribers need no knowledge of which system is available in which city, or whether their service provider covers a particular area; the phone will hunt for the most appropriate system.

In addition, intelligent roaming allows the operation of the phone to be simplified, since the phone makes the decision about which system on which to camp. No user intervention is required for the subscriber to be provided with fast, cost-effective service. Figure 8.10 shows a simplified intelligent roaming algorithm.

8.9 IS-136 phone states and operations

The IS-136 phone can be in one of a number of states at any particular time. Figure 8.11 diagrams these states, although the actual states have been condensed to 12 for clarity. These states, which range from off to processing SMS to reselection, are briefly described in Sections 8.9.1–8.9.11. Please refer to the IS-136 specification for a full description of phone state activity.

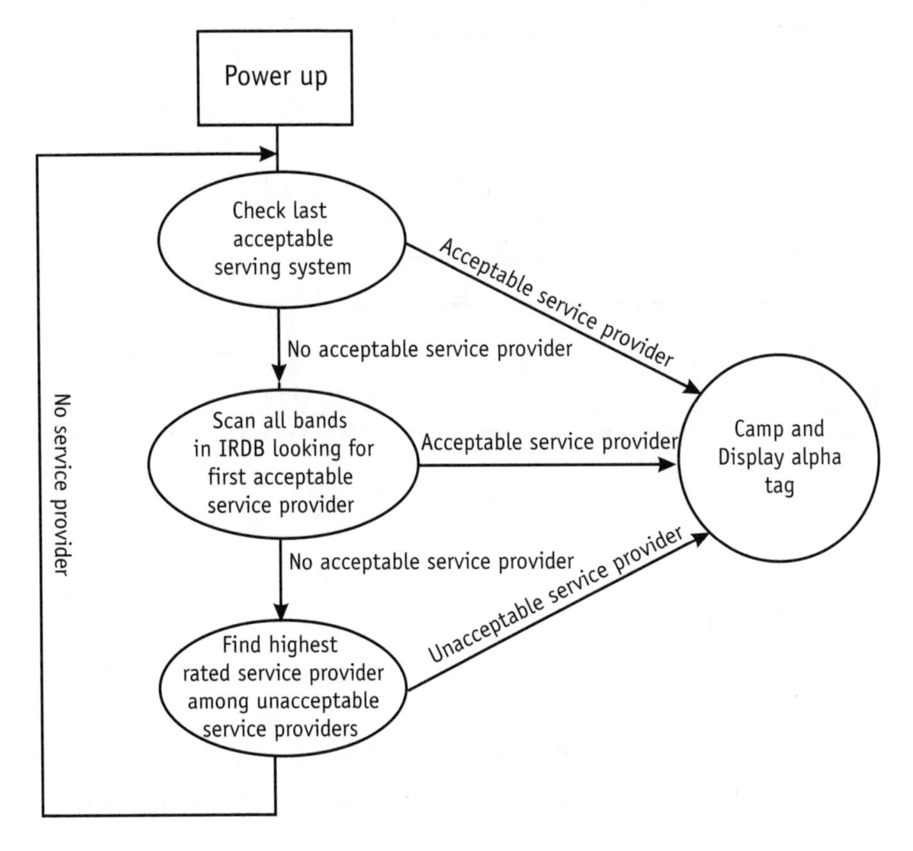

Figure 8.10 IS-136 system selection.

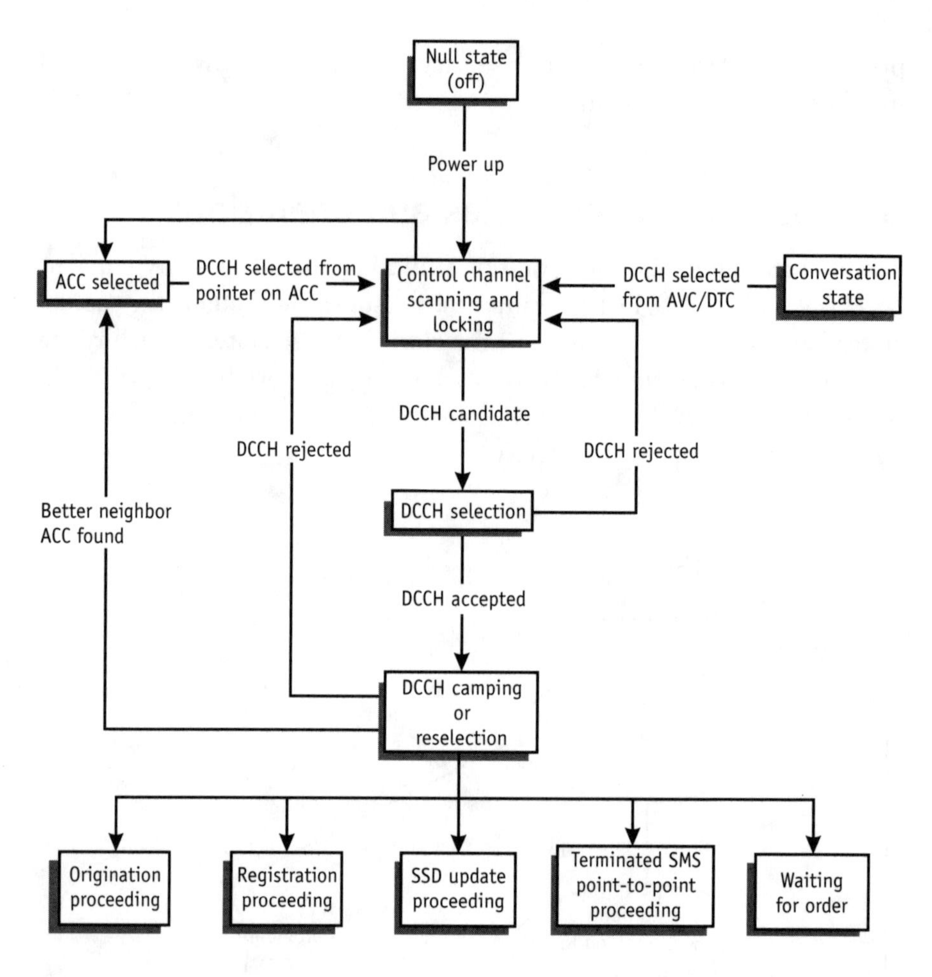

Figure 8.11 IS-136 cellular phone state diagram.

8.9.1 Null (off)

The phone is powered down; there is no phone activity in this state.

8.9.2 Control channel scanning and locking

In this state, the phone is attempting to find a suitable DCCH to gain service. This involves scanning the history list of last-used channels, investigating the ACCs for DCCH pointers, or scanning the probability blocks to find a DCCH or a DTC with pointer information. If the phone is capable of dual-band operation, this process includes intelligent roaming to determine the best band to use from the 800- and 1,900-MHz frequencies. If a likely candidate is found, the phone will enter the DCCH selection state.

8.9.3 ACC selected

In this state, the phone has found no DCCH and is selecting an ACC as the control channel. The phone can return to the DCCH environment using the DCCH pointer on the ACC if the prospective DCCH meets the camping criteria.

8.9.4 DCCH selection

The phone is evaluating the candidate channel. If the channel is suitable for gaining service, the phone will enter the DCCH camping state. If the channel is rejected (perhaps owing to poor signal strength) the phone will go back to the scanning and locking state in order to evaluate further channels.

8.9.5 DCCH camping

This is the phone's idle state, but it will be performing many tasks during this time. Reselections will be taking place in which the phone will be reevaluating the DCCH environment in order to find the optimum channel on which to camp. In addition, the phone will be responding to internal and external messages to start call processing and registrations or to receive teleservice messages.

8.9.6 Conversation

The phone is in this state when it is tuned to a voice or traffic channel and is providing a voice path for a call. The phone will be measuring MAHO neighbors (both 800- and 1,900-MHz) and responding to hand-off commands from the system by retuning to the appropriate channel. When the conversation ends, the phone will enter the control channel scanning and locking state. The call release messages on the analog voice and DTCs contain DCCH pointer information that indicates the control channel to scan for DCCH operation.

8.9.7 Origination proceeding

In this state, the phone is in the process of starting a call after the user has dialed a number and pressed send. The phone has sent an origination message to the system and is waiting for the response that indicates the digital traffic or analog voice designation for that call. When this response message is received, the phone will leave the DCCH and retune to a traffic or voice channel to commence the conversation. Figure 8.12 depicts the origination proceeding process.

8.9.8 Registration proceeding

The phone will be in this state when it has sent a registration message but has not yet received a registration-accept message from the system. Figure 8.13 depicts the registration proceeding process.

8.9.9 SSD update proceeding

The phone will be in this state during the authentication process, when it has received an authentication challenge and is waiting to submit further authentication information to the system.

8.9.10 Terminated SMS pt-to-pt proceeding

The phone will be in this state when it has sent a SPACH confirmation (similar to a page response for voice calls) and is waiting for the system to deliver teleservice information—for example, a CMT short message,

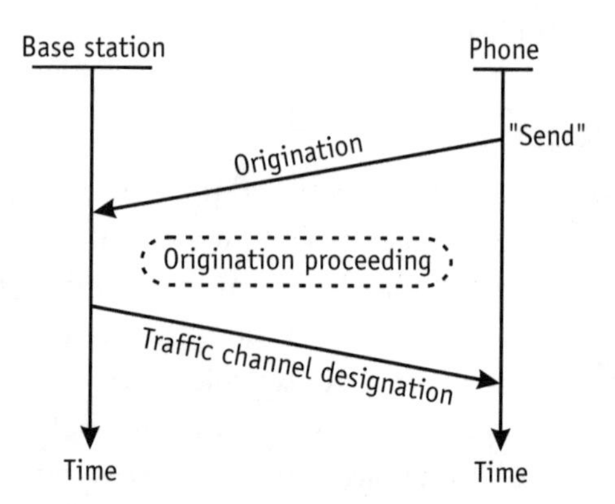

Figure 8.12 Origination proceeding state.

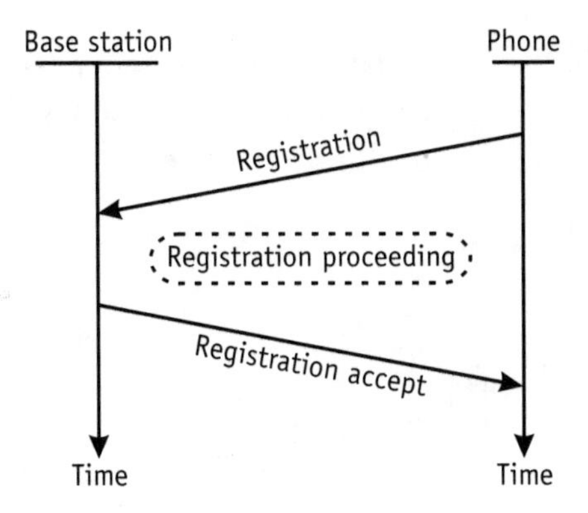

Figure 8.13 Registration proceeding state.

over the air activation information, or UDP data. The R-data message contains the actual application data, as described in Chapter 3. Figure 8.14 depicts the terminated SMS pt-to-pt proceeding process.

8.9.11 Waiting for order

In this state, the phone has received a page message, indicating an incoming call, and responded appropriately. The phone is now waiting for the

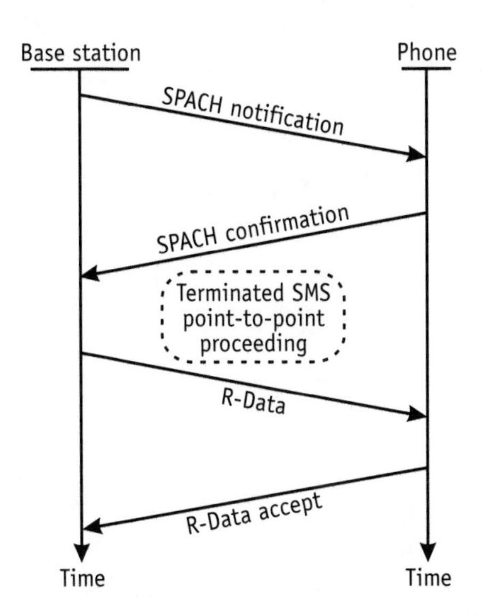

Figure 8.14 Terminated SMS Pt-to-Pt proceeding state.

traffic or voice channel designation to commence the voice call. When this message is received, the phone will leave the DCCH and retune to a traffic or voice channel for the duration of the call. Figure 8.15 depicts the waiting for order process.

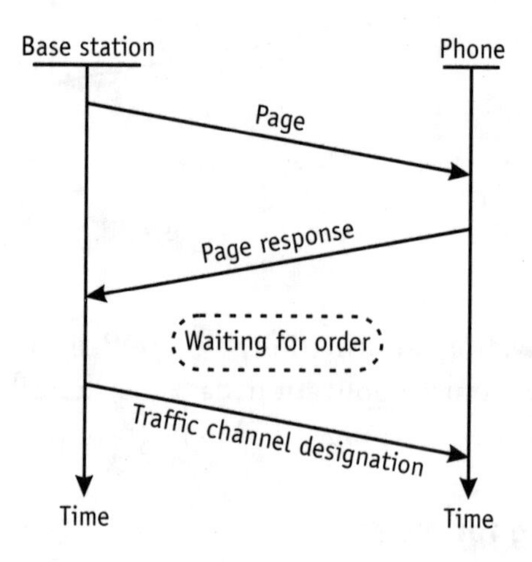

Figure 8.15 Waiting for order state.

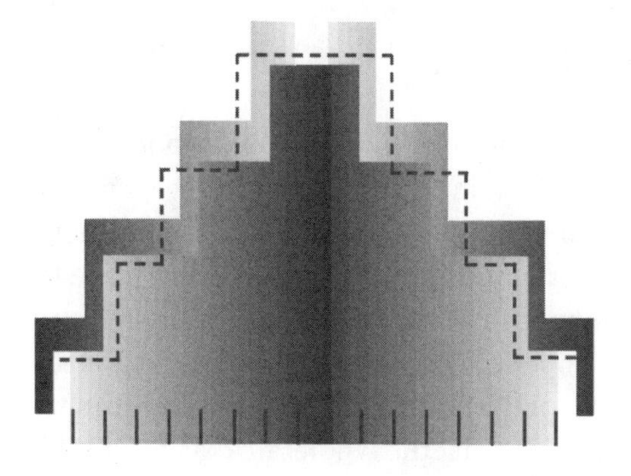

9

TDMA Economics

TDMA ECONOMICS INCLUDES the mobile phone cost, the system cost, and operational costs. In general, equipment and operations costs continue to decline as high demand for the devices which permitted mass production that reduces the average cost per device (mobile phones and system equipment). This trend is likely to continue, and increased competition from new PCS companies is likely to decrease the cost of devices and services to consumers. The costs described in this chapter are examples of general costs in the cellular industry and should not be associated with any specific equipment manufacturers or service providers.

9.1 Mobile equipment costs

The average wholesale cost of analog mobile phones has dropped from $307 in 1992 to approximately $104 in 1996.[1] The average cost of TDMA mobile phones is approximately double the cost of equivalent analog phones. The high cost of digital mobile phones is primarily due to the following factors: development cost, production cost, patent royalty cost, marketing, and post-sales support. The wholesale equipment costs of TDMA mobile phones have dropped by over 80% since their commercial introduction in 1992, and it is likely that the wholesale cost of TDMA phones will continue to decrease by more than 15% per year until they eventually equal the cost of equivalent analog phones. The cost reduction of TDMA phones has been made possible by large-scale production and the use and cost reduction of integrated circuits.

9.1.1 Nonrecurring engineering (NRE) cost

Nonrecurring engineering (NRE) costs are associated with the research, design, test, and development of a new product. NRE costs for TDMA phones can be high due to the added complexity of digital cellular phones. The proportion of NRE costs per phone depends mainly on the amount of production over which the cost is distributed. That is why the average cost of phones continues to drop each year as production volumes increase.

The NRE cost per phone can vary from $500,000 for a product that has only minor modifications from an existing product to over $10 million for a completely designed new product. Figure 9.1 shows the amount of NRE development costs per phone as the quantity varies from 10,000 to 500,000 units. Even small development costs become a significant challenge if the volume of production of the digital mobile phones is low (below 20,000 units). At this small volume, NRE costs will be a significant percentage of the wholesale price.

Typical product development costs include market research; technical trials and evaluations; industrial, electrical, and software design; prototyping; product and FCC testing; creation of packaging, brochures,

1. "The Retail Market of Cellular/PCS Telephones," *Data Flash,* Herschel Shosteck Associates, Washington, D.C., December 1996, Vol. 14, No. 1.

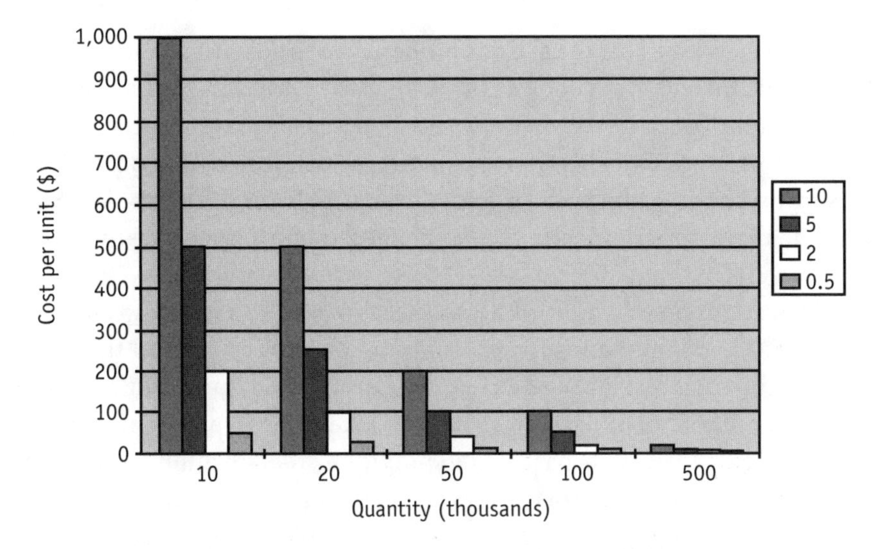

Figure 9.1 Mobile phone NRE cost.

and user and service manuals; marketing promotion; sales and customer service training; industry standards participation; unique test equipment development; plastics tooling; special production equipment fabrication; and overall project coordination.

Many design tradeoffs accompany the creation of a product. Such tradeoffs include a higher cost per unit for a quicker development period or simplified designs. Rapid development may involve outsourcing development of assemblies and rush development fees.

The most cost-effective designs for mobile phones involve a balance between integrating multiple assemblies into a custom-integrated circuit and the use of low-cost components. The use of custom-integrated circuits reduces the number of components contained by a phone. Unfortunately, custom-integrated circuits can cost more than $100K each to develop, and there may be several custom-integrated circuits in each phone. Low cost parts may decrease the cost in small quantities of production, but they will increase the size and weight of the phone.

9.1.2 Production cost

The production cost is determined by the parts cost and the automated assembly and manual assembly costs. Each phone has a bill of materials (parts list), and the cost of each component varies depending on the quantity purchased and the specification of the part. Parts also have a lead time for availability because the manufacturer of the part may have a scheduled cycle for production. This is especially important for custom parts. Rush fees can sometimes be paid to expedite the delivery of parts. In 1998, the bill of materials (parts) for a digital mobile phone was approximately $100.[2]

The circuit boards for mobile phones typically use *surface-mount technology* (SMT) components. These components are attached to the board using automated assembly equipment. Automated equipment assembly lines cost millions of dollars, and each line takes time to program, connect together, and load with parts. Depending on the assembly equipment (more modern equipment is faster) and number of components on a circuit board, a single production line can assemble from several hundred thousand to over one million circuit boards per year.

Not all assembly steps can be automated, however. Some manual assembly may be required for unusual components such as keypads, displays, or bulky radio filters. The typical average loaded cost for labor (wages, taxes, and benefits) varies from approximately $20 to $40 per hour.[3] The amount of time to assemble a phone (all nonautomated steps) can vary from about two-tenths of an hour to 1.5 hours. The average amount of time takes into account all workers involved with the plant. This includes assemblers, administrative personnel, and plant managers.

9.1.3 Patent licensing

An often neglected cost of mobile phones is the cost factor associated with patent royalties. Large manufacturing companies typically exchange the right to use their patented technology with other companies that have patented technology they want to use. Manufacturers that do not exchange patent rights must pay for the right to use the technology. Patent licensing costs can range from pennies per unit to several dollars per unit, and several patents may be required.

---

2. Personal interview, cellular telephone manufacturing manager, industry expert, April 6, 1998.

3. Personal interview, manufacturing manager, industry expert, March 13, 1998.

Companies that do not research for potential patent infringement may infringe on a patent after they have started production, and companies are not required to license their patents. As a result, if a company determined that a competitor had infringed on one of its patents, it could seek an injunction that would prohibit its rival's manufacture of the product. Moreover, if a company knowingly violates a patent, it may have to pay additional damages to the patent license holder.

9.1.4 Marketing cost

The marketing costs for manufacturers to promote their products include a sales staff, manufacturer's representatives, advertising, and trade shows. Mobile phone manufacturers typically boast a staff of well-paid salespeople who call on key customers (direct sales). They also employ technical salespeople to respond to technical issues.

Manufacturer's representatives may be used to sell cellular and PCS products. These representatives typically receive up to four percent of sales for their services.

Advertising promotion programs include brand promotion and targeted advertisements for specific products. The budget for brand recognition advertising typically ranges from less than one percent to over four percent of sales and is increasing.

Targeted advertising may be termed *pull* or *push advertising*. Pull advertising occurs when the manufacturer directly promotes its product through the mass media, and push advertising typically comes in the form of cooperative advertising.

To promote their products through retailers, manufacturers offer cooperative advertising allowances that range from approximately 2–4% of the sales amount. Retailers receive the cooperative advertising funds when they demonstrate that they have paid for local advertisements that have featured the manufacturers' product.

Manufacturers promote their products at approximately three to four trade shows each year. Trade show costs are high since trade show space costs approximately $20–$40 per square foot and manufacturers may lease hundreds of square feet for custom displays. In addition, manufacturers may provide gifts and hospitality entertainment parties to lure potential buyers to the booth and may staff dozens of people at each show to answer customers' sales inquiries. In sum, manufacturers' marketing programs cost more than 10–15% percent of their products' wholesale selling price.

9.2 System equipment costs

System equipment costs have been decreasing due to competition and economies of scale. The cost for system equipment stems primarily from the development cost, the production cost, the patent royalty cost, and the cost of marketing and post-sales support.

9.2.1 Nonrecurring engineering costs

The cost to develop system equipment is much higher than the cost to develop mobile phones. This is because the complexity of a system is significantly greater than that of a mobile phone, and the necessary validation testing is lengthy and critical to the operation of the system.

There are many more assemblies in an MSC than in a mobile phone. These include communications controllers, radio transceivers, switching assemblies, computer databases, and very complex software.

Wireless systems must be extremely reliable. All changes or additions to the system must be thoroughly tested since even a small change can affect the entire system. Testing software for wireless systems requires thousands of hours of labor by highly skilled professionals.

9.2.2 Production cost

The production cost for network equipment is determined by the parts cost and the cost of automated and manual assembly. The circuit boards for network equipment typically use SMT components. There are usually several circuit boards for each piece of network equipment.

There are thousands of phones manufactured for each base station that is produced. The quantity of production is typically much lower for base stations than the production levels for mobile phones, and the assemblies are more complex. Setting up automated factory equipment for several circuit boards is time consuming. As a result, a high proportion of the production cost is absorbed by labor expenses.

Wireless networks are designed to be flexible (for the addition of new services) and to be cost-effective. Most digital systems use DSPs that can be reprogrammed. While these powerful integrated circuits increase the cost of the bill of materials, they also allow flexibility for key changes such as new messaging or radio channel signaling changes. Moreover, DSPs are standard building blocks that reduce development time. The maturity of digital technology allows cost reductions through the use of cost-effective equipment design and low-cost commercially available electronic components.

Wireless system radio equipment requires many different connectors, cumbersome RF radio parts, and bulky equipment cases. Because the insertion of unique parts is often necessary, it is not usually possible to use automatic assembly equipment. The typical cost for each network assembly is several thousand dollars, with the cost primarily determined by the production volume. Figure 9.2 shows how the production cost per unit drops dramatically from approximately $400–$1,000 per unit to $50–$125 per unit as the volume of production increases from 5,000 units per year to 40,000 units per year. The chart in Figure 9.2 assumes that the production cost for each assembly is four times that of mobile phones due to the added complexity and the use of multiple assemblies.

9.2.3 Patent licensing

Only a few manufacturers produce wireless system equipment and network equipment benefits from the use of many different patented technologies. Large manufacturers have a portfolio of patents that they commonly cross-license and trade with other manufacturers. This tends to reduce the cost of patent rights. When patent licensing is required, the patent costs are sometimes based on the wholesale price of the assemblies in which the licensed technology is used.

9.2.4 Marketing cost

To sell wireless network equipment, system equipment manufacturers employ well-trained sales staff and invest in promotion through advertising, trade shows, and industry seminars that contribute to the overall marketing cost. The sales staff consists of several specialized representatives for key clients. Because network system sales are much more

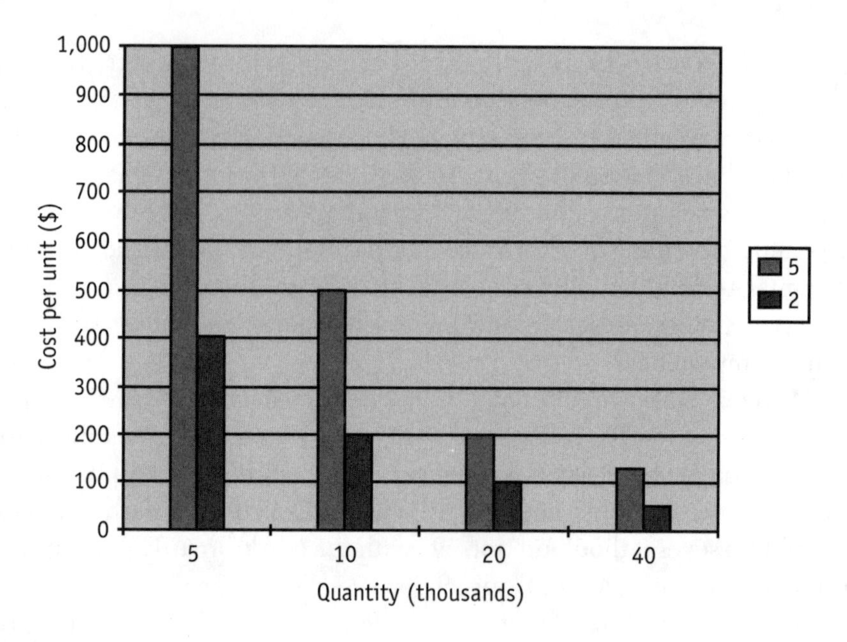

Figure 9.2 Factory assembly equipment cost.

technical than the sale of mobile phones, manufacturers typically employ skilled sales engineers to answer technical questions.

Manufacturers use broad promotion campaigns for brand recognition and advertisements targeted for specific products. The budget for brand recognition advertising has typically been small, but spending on brand recognition is increasing.

Companies often limit product advertisements to trade journals to focus on key decision makers. The sale of cellular system equipment involves only a small group of people who typically work for a wireless service provider. The advertising budget for wireless system equipment manufacturers may be less than two percent.

Much of the promotion of wireless system equipment occurs at trade shows, industry associations, and direct client presentations. System equipment manufacturers typically exhibit at three to four trade shows annually. Trade show costs for wireless system manufacturers—like those for mobile phones—can be very high. Moreover, system manufacturers often bring many more sales and engineering experts to the trade shows than mobile phone manufacturers, which increases the cost.

It is estimated that the marketing costs for wireless system equipment manufacturers are approximately 8–10% of the wholesale selling price.

9.3 Carrier network capital costs (*Source:* APDG)

Network capital cost is associated with the purchase of system equipment that services customers. The average capital cost per customer is determined by dividing the cost of the equipment by the number of customers it services. As the number of customers increases, more system equipment is typically purchased. Table 9.1 lists the estimated cell site costs.

The average capital cost for AMPS systems is approximately $1,000 per customer, and the average capital cost for a TDMA system is approximately $400 per customer. The lower cost for TDMA systems is due to the ability of the system equipment to allow several customers to simultaneously share the same equipment.[4]

There needs to be a balance between system equipment capacity and the ability to service the needs of customers. When carriers do not purchase enough equipment to service customers, blocked calls result. If carriers purchase more system equipment than necessary (excess capacity), they increase the average capital cost per customer.

Network equipment includes cell sites, base stations, switching centers, and customer databases. Digital equipment costs are typically higher than analog equipment costs. The cost savings for the carrier comes from the ability of customers to simultaneously share the base station equipment. This cost-effective capacity results when cell sites are converted from analog to digital channels, thereby allowing approximately three times the amount of customers to be served by the same cell site (up to six times for half-rate).

9.3.1 Cell site

A cell site is composed of two primary parts: a radio tower and a base station. The radio tower is typically 100–300 feet tall and typically costs

4. Personal interview, financial analyst, industry expert, March 22, 1998.

between $30–$300K. Base stations are composed of a environmentally secure building and radio equipment and cost between $100–$500K.

Cell sites are typically located on a very small area of land or on top of a building. The land is either purchased or leased, and the rights to use the top of a building are typically provided using a lease. If land is purchased, the cost can exceed $100K. If an existing tower is located in a suitable area, tower space may be leased for $500–$1,000 per month.

A secure building is required for base station radio equipment. The typical building, which must have climate control (cooling for the electronics equipment), costs $40K. In addition to the building cost, there is the site construction cost, which may be more than $50K.

It is necessary to install a communications link between the base station and the MSC. This communications line may be leased from the local phone company, or a microwave link may be used. If the communications line is leased, the estimated cost of installing a T1 communications line can be as much as $5K. If a microwave link is used, the estimated cost for the link is about $20–$30K per link.

Each base station has radio transceivers for each radio channel and controllers that coordinate the operation of the base station. The average cost of the radio transceivers (with the proportion of the controllers) is approximately $15K each. Although each cell site can have over 50 transceivers, the average number of transceivers is approximately 20 per cell site.

9.3.2 Mobile switching center

Cellular and PCS systems must contain one or more central switches or lease switching from a neighboring system. The estimated cost per customer for switching facilities is approximately $25.

The MSC is typically located near an LEC's central office switching center. The building for the MSC, which contains the switching and communications equipment, must have air conditioning and backup power supplies. The customer database is typically located in the MSC building.

In addition to the purchase of an MSC, software upgrades and special software features can be purchased at additional cost. These software upgrades and features can be several hundred thousand dollars each.

Table 9.1
Estimated Cell Site Costs. (*Source:* APDG Research.)

Item	Cost (000's)
Radio tower	$70
Building	$40
Land	$100
Install comm line	$5
Construction	$50
Antennas	$10
Backup power supply	$10
Radio channels (20 @ $15K)	$300
Total	$585

9.4 Carrier operational costs (*Source:* APDG)

The costs of running wireless system equipment include administrative staff, leasing facilities, tariffs, billing service, fraud, and marketing promotion.

9.4.1 Administration

Carriers' staffing requirements include executives, managers, marketing communications personnel, engineers, salespeople, customer service workers, technicians, and legal, finance, administrative, and other personnel to support business functions. Typical staffing for wireless service companies is approximately 2.5 employees for each 1,000 customers. If the average total cost per employee (which covers salary, expenses, benefits, and facility costs) is $40,000, this results in a monthly staffing cost of $8.33 per customer (or $40,000 per year times 2.5 employees divided by 1,000 customers divided by 12 months).[5]

5. Personal interview, telecommunications analyst, industry expert, April 12, 1998.

9.4.2 Communications lines

The typical cost of leasing a 24-channel T1 line between a base station and MSC is approximately $500 per month.[6] Because each base station can serve up to 1,000–3,000 customers, the average cost of the leased lines is typically less than 50 cents per month per customer.

9.4.3 Interconnection to the PSTN

Wireless networks must be interconnected with the LEC. The LEC typically charges a small monthly fee and several cents per minute (ranging from approximately 0.75–4 cents per minute) for each line connected to the wireless carrier. Recently, the FCC ruled that LECs must pay other networks (such as a wireless network) for calls the LEC delivers into that network. This practice is called *reciprocity*.

9.4.4 Long distance

A wireless service provider can directly connect to an IXC or provide long-distance services through an LEC connection. If the wireless carrier is directly connected to the IXC, it will directly pay the IXC for the long-distance services it provides. Typical direct charges for long-distance connections range from 5–10 cents per minute. If the wireless carrier connects to the IXC via the LEC, a portion of the long-distance charges will be paid to the LEC for access to the IXC through a POP connection. These long-distance interconnection tariffs can garner an additional 45% charge, which is based on the long-distance charges.

9.4.5 Billing services

Costs associated with billing include the gathering, printing, and mailing of billing information. Billing issues are becoming more complicated as advanced services are provided. Billing records are created as customers originate and receive calls. These billing records may be recorded in the customer's home system or in a visited system. Each billing record contains call details including who initiated the call, where the call was initiated, the length of the call, and how the call was terminated.

6. Personal interview, telecommunications analyst, industry expert, April 12, 1998.

Billing records can be stored in the wireless service provider's database, or they may be held in a visited system for transfer back to the home system. Billing records from wireless networks are either transferred via tape in a standard *automatic message accounting* (AMA) format or via *electronic data interchange* (EDI) format. The estimated cost for billing services is approximately $1–$3 per month.[7]

9.4.6 Fraud

Fraud, which is the unauthorized use of wireless service, may be accomplished by changing or manipulating the electronic identification information stored inside of a mobile phone or by falsely obtaining service through the normal subscription process. Fraudulent use of wireless services in the United States can represent over 4% of sales. This is a challenge for tariffs and long-distance charges, particularly international calls. The type of fraud has changed from subscription fraud, to tumbling, to cloning and is now moving back to subscription fraud.

When a person registers for wireless service using false identification, it is called subscription fraud. The fraudulent customer uses falsified documentation to meet the subscription requirements of the wireless service provider. Thinking that all the requirements have been met, the wireless service provider activates service. After the first billing period, the wireless carrier discovers that the bill is not going to be paid, and service is disconnected.

During the introduction of cellular throughout the United States, most cellular systems were not interconnected. Wireless carriers would bill roaming customers by sending the visiting customers' call information to a clearinghouse after the calls were made. This allowed roaming fraud as it reprogrammed the telephone's number, making it look like a roaming customer.

In the late 1980s, most cellular systems were interconnected. This allowed precall validation of roaming customers. To continue to gain fraudulent access to cellular systems, criminals began to duplicate the MIN and ESN of mobile phones to match a valid customer's identification information. This duplication of subscriber information is called *cloning*.

---

7. Personal interview, Michael H. Sommer, Information Technologies, March 26, 1998.

To overcome the challenge of cloning, the cellular system began to track usage and look for unusual signs. This allowed the quick detection of cloned phones. Subsequently, criminals quickly discovered *tumbling*, another way to circumvent this system. Tumbling is the continuous changing of MIN and ESN information in a phone to a list of valid customers.

To overcome the ability to tumble valid customer identity information, the new TDMA standard uses authentication. Authentication employs secret information and call counters that make it very difficult to gain unauthorized service. This virtually eliminates the possibility of cloning or tumbling. Now, fraudulent activity seems to be moving back toward subscription fraud. In any case, fraud will continue to be a significant cost for cellular and PCS carriers.

9.4.7 Marketing

The marketing costs for wireless carriers include a direct sales staff, agents, advertising, and activation subsidy.

Wireless carriers typically have a sales staff that services large corporate customers. The sales staff typically consists of well-paid sales leaders who have been trained for the sale of wireless products and services.

Agents include master agents and independent agents. Master agents are typically offered activation commissions and a percentage of a customer's phone bill (called *residuals*) for a period of one to three years. Independent agents typically receive an activation commission.

Advertising costs include fees to use brand names and brand promotion. It is expected that spending for advertising will increase as wireless service competition continues to rise. Some carriers may receive cooperative marketing funds from manufacturers for the distribution of their equipment.

The cost of mobile equipment is different for consumers than for distributors. Typically, consumers pay a retail price that is higher than the wholesale price. However, to entice the customer to purchase a mobile phone, the service provider will typically pay the distributor an *activation commission*. This subsidy allows the distributor to sell below the wholesale price.

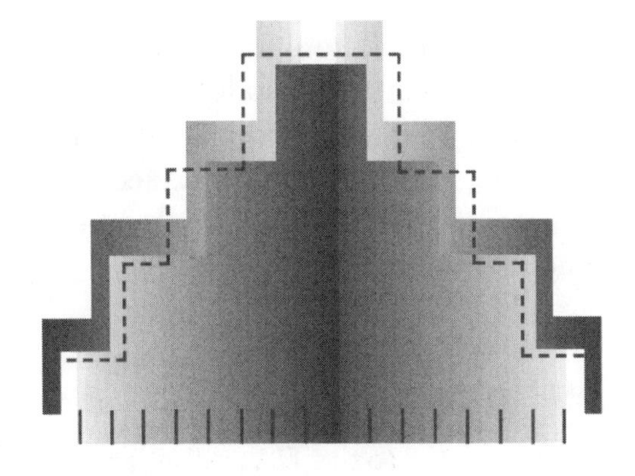

10

IS-136 Features

10.1 Introduction

There are many new features and services that make IS-136 TDMA an attractive technology for consumers. The core of many of the new features is the new DCCH, which is part of the IS-136 specification. It is the primary enhancement to IS-54B (ANSI standard TIA/EIA 627) technology and represents the next generation of TDMA-based digital-cellular operation. The DCCH was designed to provide the advanced features described in this chapter and to be able to support future services in a timely and competitive manner.

IS-136 makes TDMA a more powerful digital cellular and PCS technology by combining the advanced features of the DCCH with the backward compatibility of coexistent digital and analog networks. This means that a carrier can have both IS-136 and AMPS running side-by-side in the same network—providing basic cellular service in fringe areas and advanced service in the system core while a planned upgrade from AMPS to IS-136 takes place. In addition, IS-136 is designed to work on either the 800-MHz or 1,900-MHz frequency band, making it a true PCS technology. In this way, customers are able to receive the same advanced features and benefits regardless of the frequency used in their location.

10.2 History of the features

The IS-136 DCCH specification was presented to the TIA and adopted in November 1994. Revision A of the IS-136 specification, dealing with some of the advanced features in the original DCCH platform, was published in October 1996. Addendum number 1 to IS-136, which contained OATS, was published in November 1996. Since then, standards development has been ongoing, and an updated version (revision B) of the IS-136 specification will soon be released dealing with some additional advanced features for the DCCH platform.

10.3 Migration from AMPS to IS-136

The IS-136 specification, while encompassing all the service requirements that are driving this new technology, was also intended to simplify new product design and minimize network impact. To this end, IS-136 was designed to fit seamlessly into the existing radio and network architecture of an AMPS or IS-54B system. The DCCH takes up the same bandwidth and uses the same modulation scheme and basic call control principles (such as in-call voice handling, hand-off, authentication, *calling number identification* (CNI), and *message waiting indicator* (MWI)) as a DTC while enabling the introduction of many new key features and services.

Introducing the DCCH into an existing system is made easier because the new system maintains many of the physical radio-interface attributes from IS-54B. Although a smooth migration can be provided from analog cellular to a TDMA/DCCH-capable system, a new phone is required to take advantage of the advanced services in the DCCH package.

10.4 Service features

The IS-136 standard provides all the features included in the IS-54B system and offers many new features. Table 10.1 summarizes the key customer features of the IS-136 cellular system.

Table 10.1
Key Customer Feature List

Feature	Benefit
Extended battery life	An IS-136 phone battery can last for several days rather than hours between recharges
Dual cellular and PCS capability	Allows the customer to access either 800-MHz or 1,900-MHz systems with exactly the same features.
Access to AMPS system	Allows the customer to roam on the universally available AMPS system if no digital system can be found
Improved voice quality	Consistent improved voice quality with support for state-of the-art vocoder
Messaging services	Allows advanced services such as alpha-paging and wireless e-mail
OAA	Allows the phones to be programmed automatically over the air
Intelligent roaming	Fast acquisition of the most suitable frequency band and carrier in a multisystem environment
Private systems	Allows for wireless office service with advanced features
Best radio channel selection (reselection)	Allows the mobile phone to continuously and automatically select the most suitable serving channel
Multiple system type support (public, private, and residential)	Allows the customer to use a single phone for home, office, and macrocellular network
Continuous data (circuit-switched) and fax services	Provides a reliable data transfer link and data compression service to increase data transfer rate
Improved security and voice privacy	Reduces fraud and keeps conversations secure

10.4.1 Extended battery life

Battery life is determined by the amount of power consumed in transmit mode (talk time) and receive mode (standby time). Talk time is increased in the IS-136 by burst transmission and discontinuous transmission. Standby time is primarily influenced by the percentage of time the mobile phone can sleep compared to the time it is awake. In addition, the power saving features of the IS-136 specification and the use of low voltage integrated circuits will also extend the battery life by reducing the amount of current consumption in both the transmit and standby modes.

The new DCCH in the IS-136 specification provides for sleep mode in which phones can turn off much of their circuitry until they need to wake up, at predetermined intervals, to receive system messaging. This feature greatly increases the standby time of cellular phones, thereby increasing battery life. Currently, phone battery life has been increased from hours to days and may well last over a week in the near future. Standby time for IS-136 telephones is dramatically increased through the use of sleep modes (paging classes).

10.4.2 Dual cellular and PCS operation

Throughout large geographic regions, wireless service providers may own or have relationships with several other cellular and PCS systems for roaming. To take advantage of these systems, IS-136 defines that a mobile phone may operate in both the 800-MHz cellular and the 1,900-MHz PCS bands. This is an option defined in the standard, and both single-band and dual-band IS-136 phones are now on the market. Dual-band mobile phones must be capable of transmitting and receiving on both sets of frequencies. Dual-band customers experience the same features when operating on either the 800-MHz cellular or 1,900-MHz so-called PCS system. This allows seamless cellular and PCS operation throughout a large geographic area.

10.4.3 Capacity increase using dual-band overlays

In addition to providing feature portability for large areas, dual-band capabilities can be used to increase the overall network capacity in a defined area. The operator may have an existing 800-MHz TDMA network and overlay it with a 1,900-MHz network. Both networks can be switched out of the same MSC. Dual-band phones automatically take advantage of either the 800- or 1,900-MHz resources in a seamless fashion. This relieves demand and thereby increases capacity, on the cumulative TDMA network.

10.4.4 Access to existing analog system (universal service availability)

All TDMA phones have been dual-mode—that is, capable of communication using either the analog voice channels of an AMPS system or the DTCs of an IS-54B system. IS-136 extends that concept a step further and specifies the phone's operation to be on either AMPS analog control or IS-136 DCCH control. In this way, the phone will utilize the DCCH where available or revert to the analog control channel if no DCCH has been installed.

During a migration from AMPS to TDMA, there may be areas of existing coverage that have not been upgraded to digital radios at the fringe of the system. The dual-mode IS-136 phones will provide all the DCCH features in the digital area and support basic voice call processing in the rural coverage areas or in other areas where IS-136 is not deployed.

CNI, MWI, voice privacy, and authentication (to inhibit cloning) and other IS-54B features are all supported in the IS-136 environment to provide a seamless service offering to users. It is interesting to note that many AMPS networks can support IS-54B features if the operator decides to do so. This capability is possible because many AMPS networks are built on the same switching platforms that support TDMA, and a simple software upgrade to the network will activate IS-54B features without requiring a DTC. In a similar fashion, IS-136 networks that are not fully built out with DCCHs everywhere can still support IS-54B features on IS-136 phones when in an analog coverage area.

10.4.5 Improved security voice quality

IS-136 can support either the original IS-54B VSELP speech coder or the newly standardized IS-641 EFR speech codec. The EFR provides voice quality comparable to the landline reference (ADPCM) under normal radio channel conditions. Additionally, the EFR errored channel performance results in significant voice quality improvements.

10.4.6 Messaging services (cellular messaging teleservice)

Being able to deliver alphanumeric messages to a cellular phone is an important feature of the DCCH. The DCCH specification allows for sending messages to, and from, a phone with a variety of attributes controlling the delivery, storage, and display behavior of the messages. The messages are sent and received via a message center, which is a new node on the cellular network. This point-to-point message process uses the DCCH and the DTC for idle mode and in-call messaging, respectively.

Examples of point-to-point messaging include standard numeric paging, alphanumeric paging, and notifying a user about new e-mail messages. Message lengths of up to 239 characters are supported.

IS-136 also supports phone-originated (uplink) messaging, and IS-136 revision B will include a broadcast capability to multiple users. Examples of SMS broadcast messages are transmitting traffic or weather information to user groups. Figure 10.1 shows a SMS scheme.

A teleservice is a feature for transferring application data to and from cellular phones. This includes the CMT, which delivers short alphanumeric messages to the phone, and the OATS, which allows for delivery of provisioning information to the phone.

10.4.7 Over-the-air activation

OAA simplifies the phone activation process for both the subscriber and service provider by providing the capability for delivering NAM information and updates to the mobile station over the air without the need for "manual" intervention or programming of the phone. This means that the phone number and other identities are automatically programmed using messages from the system rather than laborious keypad input.

10.4.8 Intelligent roaming

In a traditional cellular environment in the 800-MHz frequency band, there were originally two cellular providers—the nonwireline and wireline service provider operating the A and B band, respectively. With the popularity of PCS services at 1,900 MHz, there are now many (up to eight, two cellular and six PCS) service providers in any given geographical area, some with and some without compatible air interface technologies or roaming abilities.

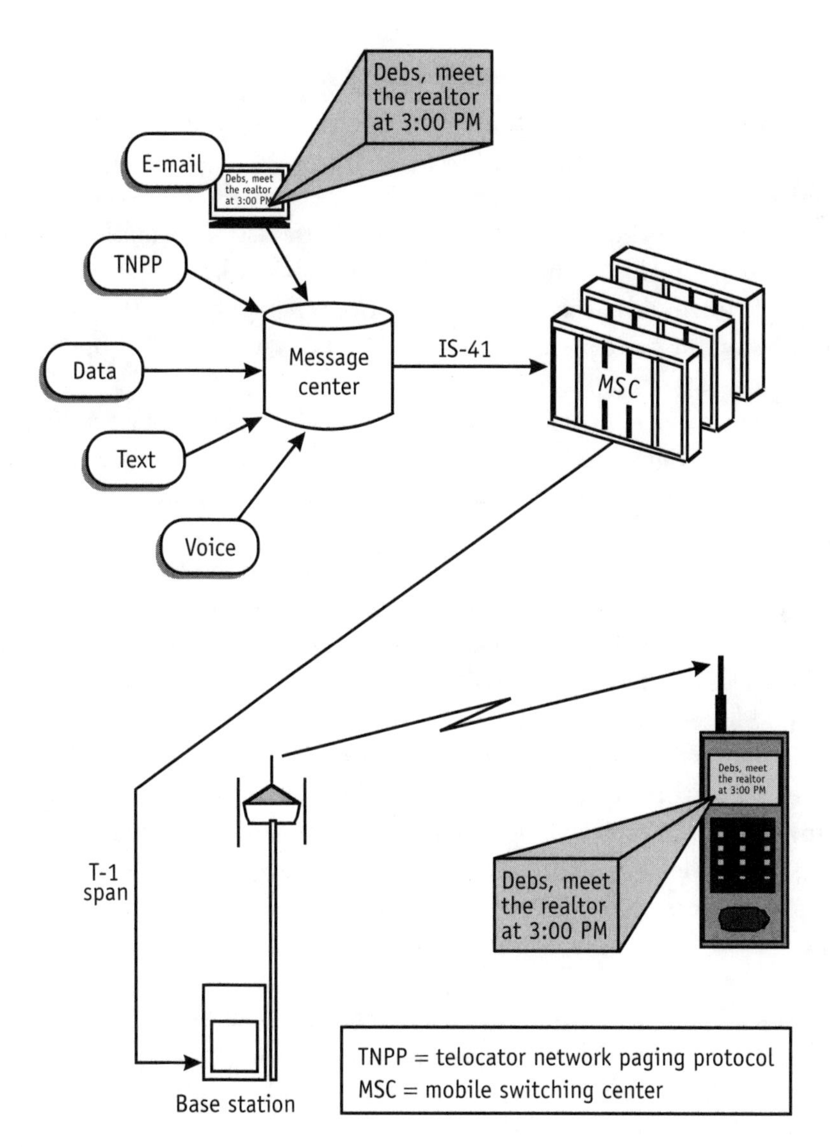

Figure 10.1 Messaging services.

Intelligent roaming is a process that allows the phone to seek out the preferred service provider from all the other sources of RF energy in the two 800- and six 1,900-MHz frequency bands. Again, the information is programmed over the air using IS-136 messages to download a database of service providers to the phone.

10.4.9 Private systems and wireless office service

The DCCH introduces features to facilitate the creation of private systems. A private system is a cell or cells that provide special services to a predetermined group of subscribers. Private systems are used in campus environments, office parks, or hospitals. In the latter example, doctors and nurses could have four-digit dialing and a special billing rate inside the hospital but would revert to regular air-time charges outside the hospital. In this way, the same phone can be used for both wireless PBX operations and as a regular cell phone.

The features to provide this capability include private-system identifiers for marking specific base stations as part of a private system, hierarchical cell structures for defining cell preferences, and new registration features to complement private systems.

A key characteristic of the private system is that the phone has knowledge of the private system and can "prefer" those channels. This is important since the public system can be stronger than the private system in certain areas, but the phone will stay on the private system as long as an acceptable service is provided.

Private systems also have tremendous flexibility for the operator since they can be created as part of the public cell sites (virtual private system) or created by installing a standalone microcellular private system. These cells then have a dual role—to provide regular wireless coverage for public users and to provide enhanced services for members of the private system. In this way, multiple services can be provided from the same infrastructure.

In a virtual private system, traffic channels can be shared between the systems. A virtual private system is useful in applications in which the subscriber base on the private system does not justify the installation of a standalone system or where there are both public and private system users in the same area (as in hospitals where doctors and nurses may be members of a private system, but visitors would use the public system).

10.4.10 Alphanumeric system ID

Alphanumeric SID or alpha tag is a new DCCH feature that places a system banner or company name on the display of the phone (shown in Figure 10.2) to let the user know in which network they reside. The alphanumeric SID is transmitted by the serving network and is available on both public and private systems. The alphanumeric SID is used to indicate to the wireless user that a billing or service difference is available in the vicinity.

10.4.11 Best radio channel selection (reselection)

DCCH cellular phones will frequently scan the surrounding control channels to determine the best channel for receiving service. This feature, which is called reselection, aids in locating microcells, private systems, and residential systems in a DCCH environment and in providing wireless office and other advanced services to the customer.

The DCCH supports the hierarchical cell concept. These enable the DCCH to identify and designate neighboring cells as regular, preferred, or nonpreferred. A DCCH phone uses that information to assess the most suitable control channel to receive service, allowing the phone to select and camp on a cell even when the phone is not receiving the highest signal strength from that cell.

For example, when a low-power microcell is providing capacity in a dense traffic area that is also served by a high-power macrocell, the HCS allows the phone to give preference to a microcell in a multitier environment. Without an HCS, the cellular phones would have difficulty camping on microcells, and the cellular system would require careful parameter settings. The HCS can also be provided for hand-off on the analog voice channels and DTCs.

10.4.12 Multiple-system type support (public, private, and residential)

The DCCH introduces the ability to categorize each cell into three network types—public, private, or residential—and allows the cellular phone to react to serving cells based on the broadcast identifiers of those network types. The three network types are defined as follows:

- Public cells refer to cells in the existing cellular system.
- The private designation is used for in-building company systems with specific features.

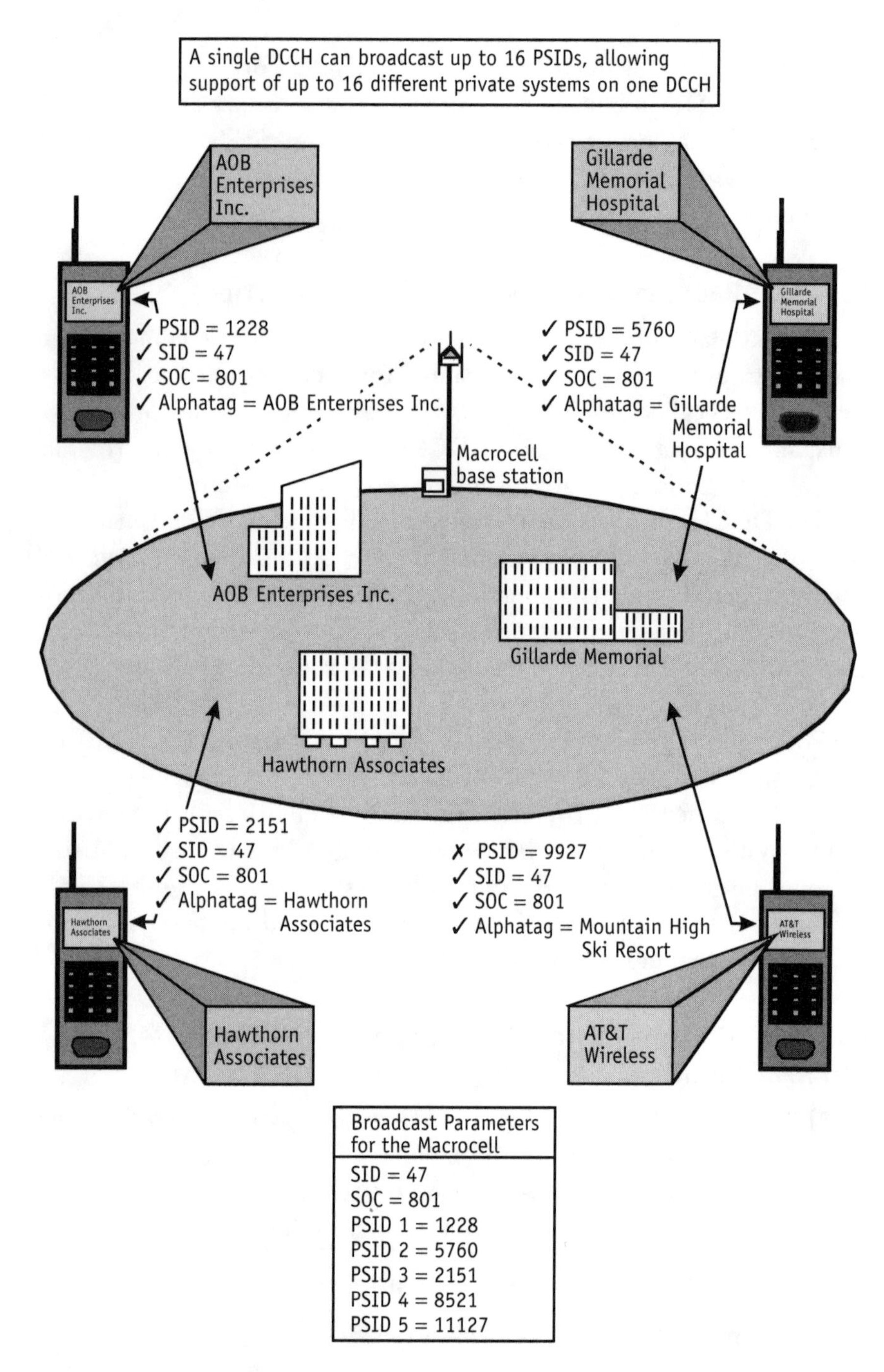

Figure 10.2 Private systems.

- Residential systems allow a cellular phone to behave like a cordless home phone.

In this manner, a single phone can act like a cordless phone at home, a wireless extension to a subscriber's desk phone at work, and a regular cellular or PCS phone while roaming on digital or analog systems.

10.4.13 Circuit-switched data services

The IS-136 specification has two companion standards, IS-130 and IS-135, that provide support for circuit-switched data in the TDMA environment. This allows asynchronous data and group-three fax transactions to take place using the same DTCs that are used for voice calls.

In addition, the radio planning, private system architecture, and channel capacity enhancements such as hierarchical cell structures and adaptive channel allocation are all maintained for both voice and data services.

Expected data rates range from a raw rate of 9.6 Kbps on a single traffic channel with no data compression to an ideal rate of 115.2 Kbps using

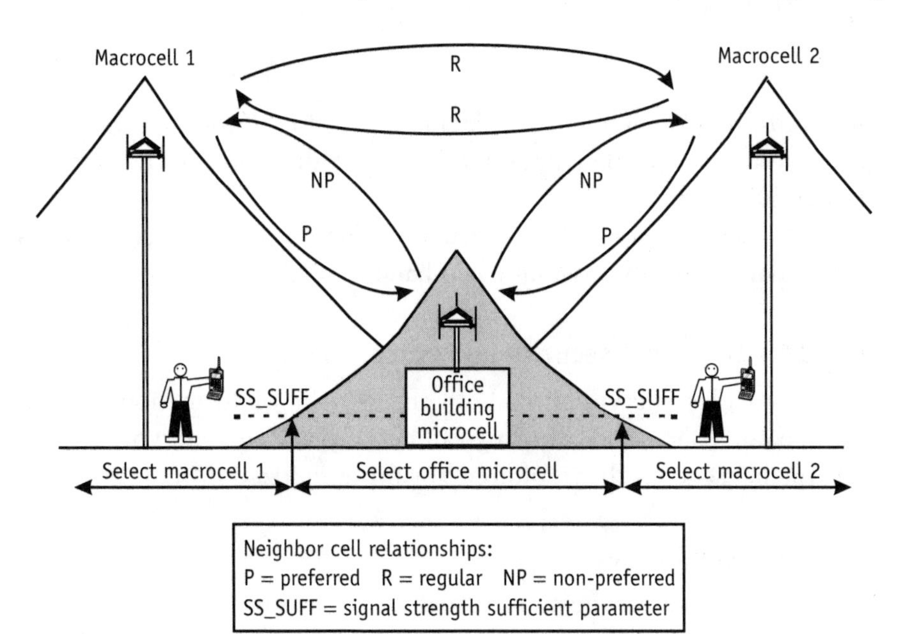

Figure 10.3 Hierarchical cell structures.

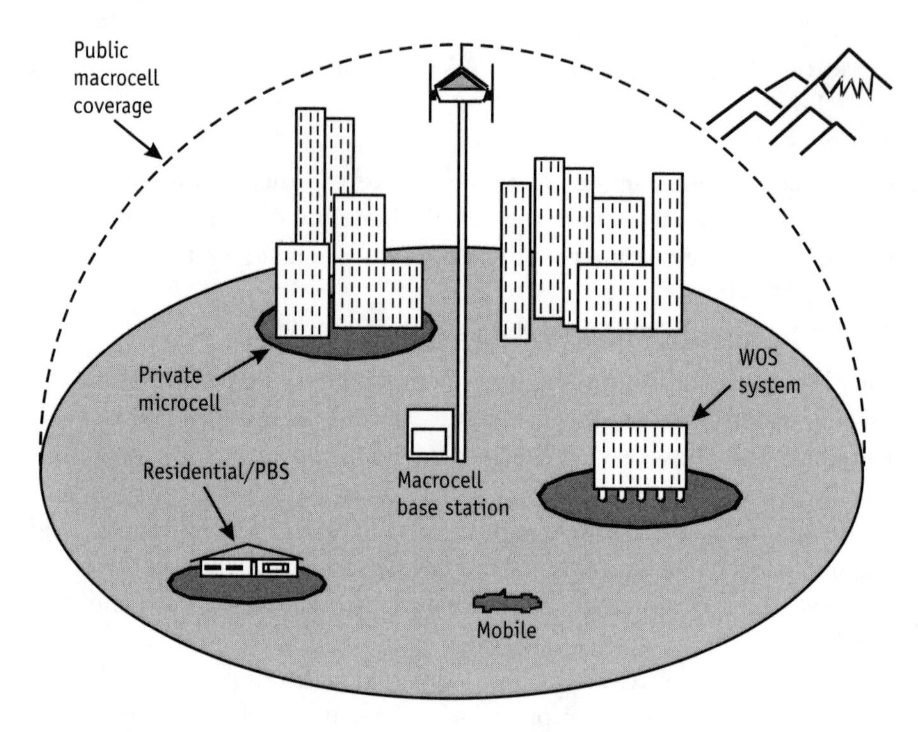

Figure 10.4 Multiple system types.

three full-rate channels and 4:1 data compression. (Note that this is an ideal case and is dependent upon such factors as the source data, compression, and RF conditions.)

IS-136 also supports circuit-switched data on analog voice channels for compatibility with existing voice-band modem technology.

10.4.14 Improved security and voice privacy

IS-136 provides encryption of both the signaling information for call setup, maintenance, and tear-down and DTC data. This encryption gives users further security than is offered by simple digitization of the traffic.

10.4.15 Time of day

The current time can be transmitted by the system and used by the mobile station to provide a clock function or to timestamp events in the mobile (such as receipt of a message).

10.5 System features

The IS-136 standard provides all the features included in the IS-54B system and offers many new features. Table 10.2 summarizes the key customer features of the IS-136 cellular system.

10.5.1 Network cost reductions

TDMA can coexist with existing AMPS radios and frequencies. This is important since in the 800-MHz band, there will most likely be existing customers still using AMPS phones and the hardware will be needed in the infrastructure to support them. A cost-effective approach is to employ common infrastructure equipment for both existing and new users. In addition, most of the infrastructure is reusable for IS-136 as are the frequency allocations in regular cellular systems.

Digital capability can be introduced when and where it is needed. When the AMPS radios are replaced to support IS-136, the AMPS equipment can be selectively upgraded or redeployed elsewhere in the system. In the meantime, digital capability can be rolled into the most important areas of the system, sector by sector.

Table 10.2
Key System Feature List

Feature	Benefit
Network cost reduction	Shared resources reduce the average network equipment cost per customer
Maximum customer capacity increase	Increases the maximum number of customers a single system can serve
Flexible roaming features	Reduces the need for customer service support for roamers
Multiple identities for mobile phones	Increases paging capacity and provides unique identity for international roamers
Design for future enhancements	Allows the carrier to implement IS-136 technology with the anticipation that new features will be easily added

10.5.2 Maximum customer capacity increase

TDMA introduces at least a threefold increase over AMPS capacity and a six-time AMPS capacity gain with the use of half-rate vocoders, and even greater gains can be achieved with hierarchical cell structures and adaptive channel allocation.

Adaptive channel allocation is a system feature that assigns the least interfered radio channels, rather than arbitrarily assigning the next available predefined channel, at a cell site. This can reduce interference, increase capacity, and provide higher quality calls.

10.5.3 Mobile assisted channel allocation

IS-136 allows the network to query mobiles as to the RF environment the mobiles are seeing. This allows the network to "sniff" areas in between and around cell sites for underutilized RF channels and re-allocate those channels to be better utilized. This is an interesting feature as it allows the "viewing area" of a cell site to be increased without adding hardware to the cell site.

10.5.4 Multiple identity structures

Private system identities, cell identities, and new system operator codes have been introduced to support DCCH features and capabilities.

The DCCH provides new flexibility in supporting cellular phone identities. The traditional use of ten-digit directory numbers for uniquely referencing cellular phones has been supplemented to allow for full international roaming. In addition to support for today's MIN, the air interface provides support for IMSIs for international roaming, and TMSIs.

A system identity structure allows cellular phones to distinguish between public, private, semiprivate (supporting both a public and private system), and residential base stations. PSIDs are broadcast so that a phone can determine whether it has special services from a particular cell when reselecting a DCCH. System operator, base station manufacturer, and country codes are also supported.

10.5.5 Design for future enhancements

The addition of a DCCH, and its associated messaging, allows a substantial amount of future-proofing to be built into the cellular air interface, enabling new products and services to be rolled out into the cellular network very quickly. This ability is key for introducing new features in a timely manner.

The IS-136 specification is also designed to support signaling specific to system operators and base station manufacturers, enabling introduction of new teleservices without the revision of the air interface.

10.6 Anticipated future IS-136 enhancements

The wireless standards are constantly evolving to offer new and improved features to the subscriber. Sections 10.6.1–10.6.7 discuss features that are under active consideration to enhance the IS-136 feature set.

10.6.1 New modulation scheme

Existing TDMA channels use $\pi/4$ DPQSK modulation. Using this modulation, with each phase change on the air interface, two bits of digital information are transmitted. New modulation schemes are under discussion to increase the throughput of the air interface but maintain the same channelization for backward compatibility. This will improve data rates for data and fax services and provide a wider pipe for more advanced voice processing technology.

10.6.2 Broadcast short message service

Currently, the teleservices in IS-136 work under a point-to-point mechanism, meaning that any teleservice message sent across the air interface can only be destined for a single mobile. If the same message is to be sent to many different mobiles, it has to be resent many times. This is obviously a less optimum use of the radio spectrum than if the single message could be addressed and received by many mobiles. The broadcast teleservice would allow such messages to be designated for all or for a subset of mobiles. Broadcast information could include traffic news, weather information, or special content that is only provided to subscribers.

10.6.3 Extended battery life in transmit mode

A discontinuous (DTX) mode that is designed to improve the talk time of phones, much like sleep mode increases the standby time, is now being considered. Under the operation of this mode, the phone would turn off unnecessary circuitry while not receiving information from the system. Comfort noise, which makes this feature transparent to the user, would also be included.

10.6.4 Packet data service

As discussed in Chapter 7, while digital circuit-switched data and fax offers an extremely versatile and reliable data and fax bearer for circuit-switched applications, a packet data solution is required to provide the efficiencies associated with applications that transmit or receive bursty information but that do not require a data connection all the time. Many options are available for packet data solutions in IS-136, and service is expected to be available in 2000.

10.6.5 Teleservice transport enhancements

In addition to a broadcast teleservice enhancement, segmentation is being proposed that would allow many teleservice messages to be "daisy-chained" together and treated as one long message at the receiving side. Associated with this capability is the assignment to traffic channels of long teleservice messages to avoid congesting the DCCH. This would increase the current limit of around 200 characters in a single teleservice message to many times that number.

10.6.6 Enhanced voice privacy and data security

IS-136 future revisions are intended to take into account new encryption algorithms to enhance the existing privacy available to customers.

10.6.7 Calling name ID

The ability to transmit the caller's name along with the caller's phone number will be supported in future releases of IS-136.

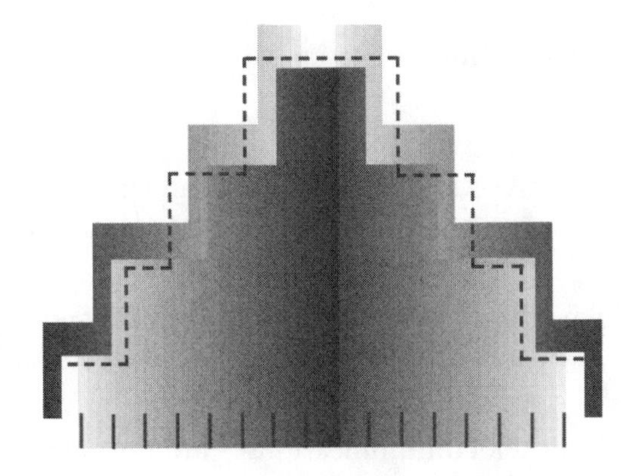

Appendix A: TDMA System Standards

TIA/EIA/IS-130—800-MHz Cellular Systems—TDMA Radio Interface—Radio Link Protocol 1.

TIA/EIA/IS-135—800-MHz Cellular Systems—TDMA Radio Interface—Async Data and FAX.

TIA/EIA/IS-136.1—800-MHz TDMA Cellular—Radio Interface—Mobile Station-Base Station Compatibility—Digital Control Channel.

TIA/EIA/IS-136.2—800-MHz TDMA Cellular—Radio Interface—Mobile Station-Base Station Compatibility—Traffic Channels and FSK Control Channel.

TIA/EIA/IS-137–800-MHz TDMA Cellular—Radio Interface— Performance Standard for Mobile Stations.

TIA/EIA/IS-138—800-MHz TDMA Cellular—Radio Interface—Minimum Performance Standard for Base Stations.

TIA/EIA/IS-641—TDMA Cellular/PCS—Radio Interface— Enhanced Full Rate Speech Codec.

TIA/EIA/IS-41.1—Cellular Radiotelecommunications Intersystem Operation: Functional Overview.

TIA/EIA/IS-41.2—Cellular Radiotelecommunications Intersystem Operation: Intersystem Handoff Information Flows.

TIA/EIA/IS-41.3—Cellular Radiotelecommunications Intersystem Operation: Automatic Roaming.

TIA/EIA/IS-41.4—Cellular Radiotelecommunications Intersystem Operation: Administration and Maintenance Information Flows and Procedures.

TIA/EIA/IS-41.5—Cellular Radiotelecommunications Intersystem Operation: Signaling Protocols.

TIA/EIA/IS-41.6—Cellular Radiotelecommunications Intersystem Operation: Signaling Procedures.

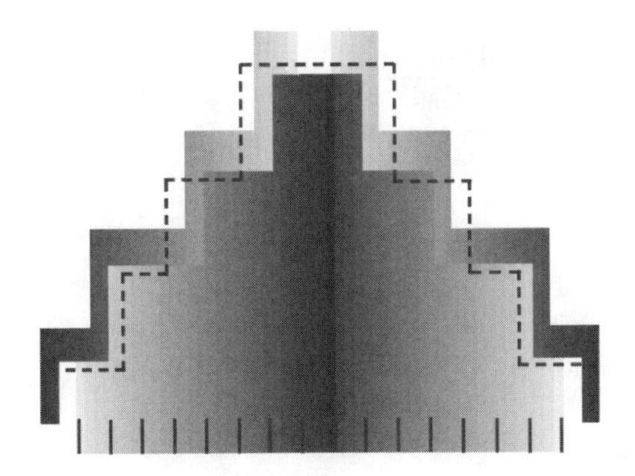

Appendix B: World Listing of Countries Using AMPS and IS-136 TDMA

THE AMPS SYSTEM is in service in the countries listed in Table B.1. IS-136 is in operation in the countries indicated.

Table B.1

Countries Using AMPS and IS-136. (*Source:* UWCC.)

Country	IS-136	Country	IS-136	Country	IS-136
Angola		Anguilla		Antigua	
Argentina	X	Aruba	X	Australia	
Bahamas	X	Bangladesh		Barbados	
Barbuda		Belize	X	Bermuda	X

Table B.1 (continued)

Country	IS-136	Country	IS-136	Country	IS-136
Bolivia	X	Botswana		Brazil	X
Brunei		Burma		Burundi	
Cambodia		Canada	X	Cayman Islands	
Central African Republic		Chile	X	China	X
Columbia	X	Congo	X	Costa Rica	X
Cuba		Curacao	X	Czech Republic	X
Dominican Republic		Ecuador	X	El Salvador	X
Gabon		Georgia		Ghana	
Grenada		Guadeloupe		Guam	
Guatemala		Guyana		Haiti	
Honduras		Hong Kong	X	Indonesia	X
Israel	X	Ivory Coast		Jamaica	
Kazakhizstan		Korea		Kurgastan	
Laos		Lebanon		Madagascar	
Malaysia	X	Maldives		Marshall Islands	
Martinique		Mexico	X	Montserrat	
Myanmar	X	Nauru		Netherlands Antilles	
Netherlands Windward Islands		New Zealand	X	Nicaragua	
North Marianas		Pakistan		Panama	X
Papua New Guinea		Paraguay	X	Peru	
Philippines		Puerto Rico	X	Russia	X
Samoa		Singapore		Sri Lanka	
St. Kitts & Nevis		St. Lucia		St. Martin / Bartholemy	
St. Vincent / Grenadines		Suriname	X	Tadjikistan	
Taiwan		Thailand		Tonga	
Trinidad & Tobago		Turkmenistan		Turks & Caicos Islands	X
Ukraine	X	United States	X	Uruguay	X

Table B.1 (continued)

Country	IS-136	Country	IS-136	Country	IS-136
Uzbekistan	X	Venezuela	X	Vietnam	X
Virgin Islands		Zaire		Zambia	

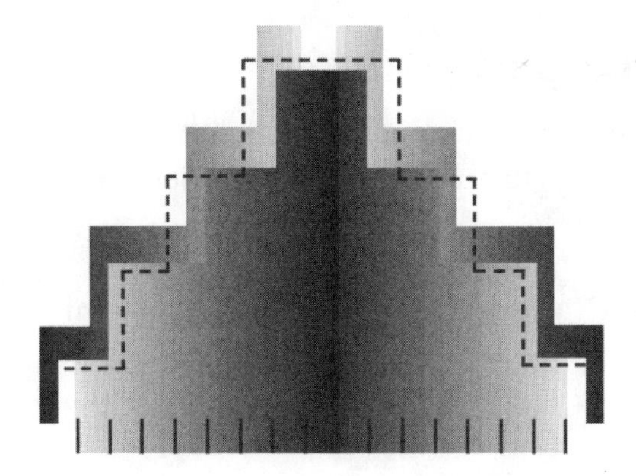

Appendix C: Control Channel Selection and Camping Criteria

After a viable DCCH candidate has been found during the DCCH scanning procedure, the phone executes a series of checks and comparisons to verify that the candidate DCCH is acceptable for camping and providing service. The operation comprises the following two procedures:

1. Signal-strength aspects determination;
2. Service-aspects determination.

Please note that this appendix describes the signal-strength (RF) aspects. The service-aspects determination ensures that the candidate DCCH supports the services that the phone requires.

C.1 Signal-strength aspects criteria

To be considered an acceptable control channel, the candidate control channel must meet the signal-strength aspects criteria described in the following equations:

$$RSS - RSS\_ACC\_MIN - MAX\,(MS\_ACC\_PWR - P, 0) > 0 \text{ dBm} \tag{C.1}$$

and

$$(MS\_ACC\_PWR \leq 4 \text{ dBm and } Mobile\_Station\_Power\_Class = 4) \tag{C.2}$$

or

$$(MS\_ACC\_PWR \geq 8 \text{ dBm}) \tag{C.3}$$

where

- RSS = received signal strength at the phone (mobile station);
- RSS_ACC_MIN = minimum received signal level required for MS to access the sector;
- MS_ACC_PWR = maximum nominal MS output power for initially accessing the network;
- P = maximum nominal MS output power defined by its power class;
- Mobile_Station_Power_Class = MS power class.

The RSS is an averaged value of the last five consecutive signal-strength measurements, with the time interval between two measurements is greater than 20 ms.

RSS_ACC_MIN, and the MS_ACC_PWR are controlled by parameters that are broadcast on the F-BCCH.

Equation C.1 means that the acceptable control-channel signal strength received at the phone shall be greater than the minimum signal strength required for accessing the base station (that is, RSS_ACC_MIN) with an adjustment owing to differences in phone transmit-power specifications (that is, MAX (MS_ACC_PWR − P, 0)). The latter term places an adjustment factor in the equation for phones that are unable to transmit at the power level specified by the base station (MS_ACC_PWR).

Equation C.2 indicates that the phone output-power level shall be within the nominal range specified in IS-136.

If the candidate DCCH does not meet these criteria, the phone will mark this cell as ineligible and attempt to find another channel. Otherwise, the candidate control channel is marked eligible, and the phone will attempt to acquire service on that DCCH.

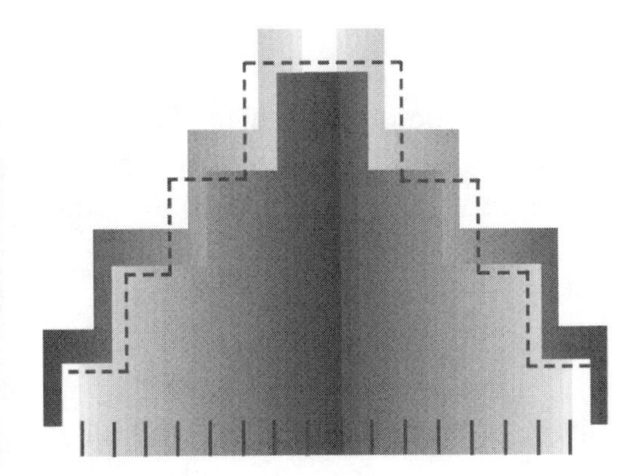

Appendix D: System Reselection

D.1 Description of reselection

Each DCCH broadcasts information that the phone uses to determine the most suitable channel to camp on and acquire service. In addition, each DCCH broadcasts information about the channels used in adjacent cells and sectors. These are called neighbors cells, and the information is broadcast on a neighbor list on the E-BCCH of the serving DCCH.

D.2 Cell threshold and hysteresis parameters

Sections D.2.1–D.2.9 describe the cell threshold and hysteresis parameters and their effects on control-channel selection and reselection.

D.2.1 Cell type

This parameter identifies the preference type of the neighbor cell. The cell type is set as either regular, preferred, or nonpreferred. The cell type is used by the phone to determine which reselection algorithm will be used in the control-channel reselection process.

D.2.2 Network type

A cell can be marked either public, private, residential, or a mix of all three, to distinguish the types of services available on particular cells.

D.2.3 RSS_ACC_MIN

This parameter indicates the minimum control-channel signal strength (received at the phone) required to access the cell (base station). Since this parameter defines the coverage area of the control channel, it can be set to the minimum received-signal-strength required for an IS-136 phone to achieve the required error-rate performance (BER or WER) on the channel serving boundaries.

D.2.4 SS_SUFF

This parameter defines the minimum signal strength considered sufficient for a control channel to be considered for control-channel reselection. SS_SUFF is the parameter used to control cell reselection during preferred and nonpreferred reselections.

D.2.5 RESEL_OFFSET

This parameter is used to increase or decrease the preference of a candidate neighbor cell being considered for cell reselection.

The RESEL_OFFSET can be set to either positive or negative values. Positive values favor reselection, whereas negative values discourage reselection. It can be compared to a hand-off bias. RESEL_OFFSET is used in all reselections.

D.2.6 MS_ACC_PWR

This parameter defines the maximum nominal-output power that an IS-136 phone can use to initially access the network.

D.2.7 SERV_SS

This parameter is a non-negative offset value that can be used as a bias in the *service offering* cell-reselection process and that is invoked autonomously by DCCH-capable phones. This parameter should be specified for each current cell.

D.2.8 DELAY

This parameter defines the length of time for which a candidate control channel must meet the required signal-strength condition in the control-channel reselection process. It is used to avoid the ping-pong effect at the cell boundaries.

The setting of this parameter is determined by the size of overlapping area between the cells, and the average mobility rate of the phones within the area. For a smaller overlapping area and/or higher average phone mobility rate, this parameter can be set shorter. For a larger overlapping area or lower average phone mobility rate, it can be set longer.

Since this parameter is time-related, it can also be used to prevent fast-moving mobiles from reselecting microcells—fast mobiles might not meet the reselection criteria for the time period specified by DELAY. This would allow only slower moving pedestrian phones to measure the microcell neighbor channels long enough to become viable reselection candidates. This would be useful when a microcell is providing service to a WOS customer situated by a freeway. The slower moving WOS customers would reselect the microcell, but vehicular traffic driving past the microcell might not meet the reselection criteria before passing out of the coverage area.

D.2.9 HL_FREQ

This parameter is used to determine the periodicity of neighbor-channel measurements.

D.3 Description of reselection algorithms

This section details the types of reselection that can occur from one DCCH to another. Reselection is the basis of hierarchical cell structures, which allow phones to choose the most suitable channel to gain service—not necessarily the channel with the highest signal strength. This is useful in microcell environments used for private systems or for traffic capacity management as shown in Figure D.1.

The algorithms described in this section use two terms to identify parameters—current and candidate, indicated as [curr] and [cand] in an algorithm.

Parameters that refer to the serving or current DCCH—that is, the DCCH on which the phone is camped—are preceded by the term [curr]. A parameter that is on the neighbor list and relates to a cell under evaluation as a new control channel—that is, a neighbor DCCH that is being evaluated by the phone to determine if it is a more suitable control channel—is preceded with the term [cand]. As previously described in Chapter 3, each cell has its own set of reselection parameters.

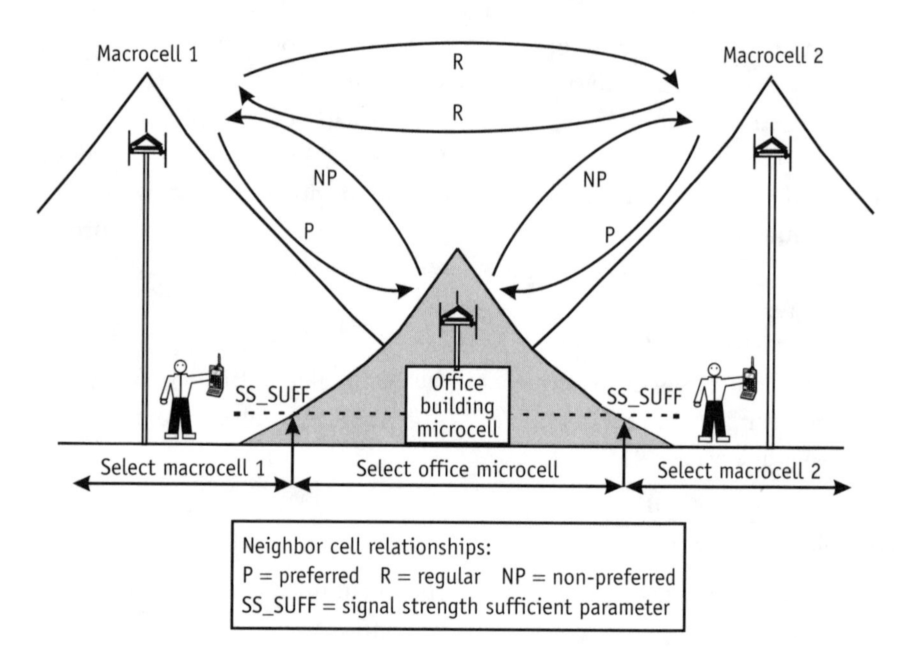

Figure D.1 Reselection example.

The following definitions also apply to the algorithms used here.

- P = The maximum mobile power output as defined by the station class mark P, therefore, it can take on the value 28, 32, and 36 dBm.
- RSS = Received signal strength at the phone (downlink).

Other factors are considered during this process (such as mobile-transmit power and signal strength above minimum threshold). These and other algorithms are described in detail later in Section D.8.

D.4 Reselection due to radio link failure

The following algorithm is used to assess a candidate control channel when the radio link from the serving control channel has failed abruptly:

$$\begin{gathered}\text{RSS [cand]} - \text{RSS\_ACC\_MIN [cand]} - \text{MAX} \\ (\text{MS\_ACC\_PWR [cand]} - \text{P, 0 dBm}) > 0\end{gathered} \tag{D.1}$$

This algorithm assesses the candidate neighbor channel for basic camping ability—that is, having a candidate signal strength above the minimum signal strength required for DCCH access on that neighbor and having mobile output power within the allowable range.

MAX (MS_ACC_PWR – P,0) will create an adjustment factor if the phone output power (P) is unable to meet the specified phone output power (MS_ACC_PWR) for that cell. This basically shrinks the cell size for low-power phones in systems designed for high-powered mobiles.

D.5 Reselection due to periodic scanning

Periodic scanning is the routine evaluation of the signal strengths of neighbor control channels—that is, channels defined on a cell's neighbor cell list. To simplify the following description, it is assumed that RSS_ACC_MIN and MS_ACC_PWR are set equally in the current and candidate cells. The equations are explained in full detail later in this section.

For each of the reselection cases, the following camping criteria should be met to ensure that the candidate's received signal strength and mobile output power are within the allowable range:

$$\text{RSS [cand]} - \text{RSS\_ACC\_MIN [cand]} - \text{MAX} \\ (\text{MS\_ACC\_PWR [cand]} - P, 0 \text{ dBm}) > 0 \qquad \text{(D.2)}$$

D.6 Neighbor candidate eligibility

The eligibility of each neighbor—that is, how suitable a particular neighbor is for reselection—will be evaluated differently for each neighbor type (either regular, preferred, or nonpreferred) as described in Figures D.2 and D.3.

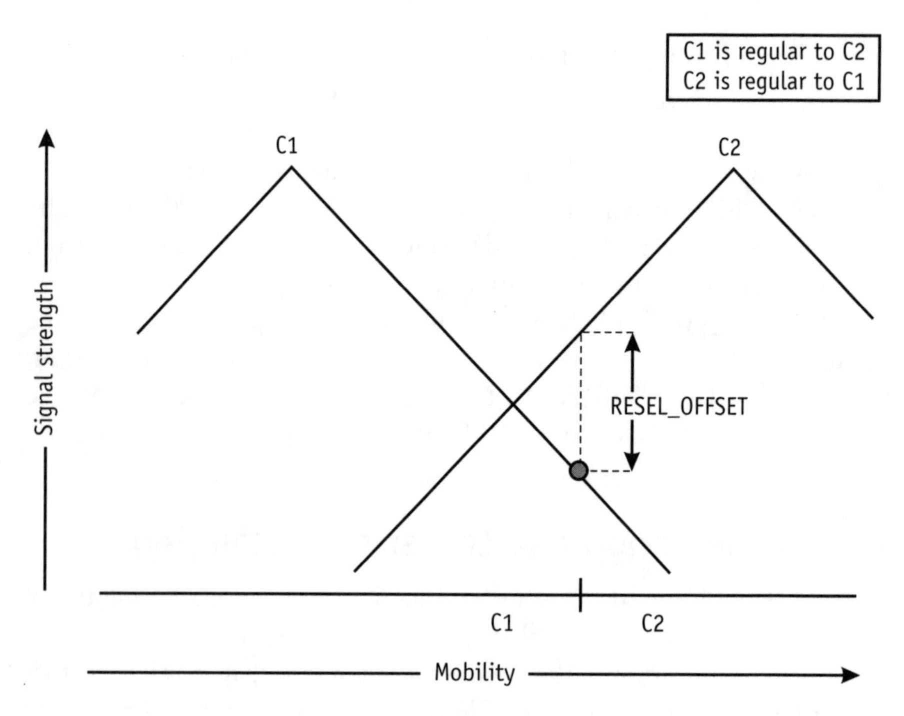

Figure D.2 Regular neighbor reselection scenario.

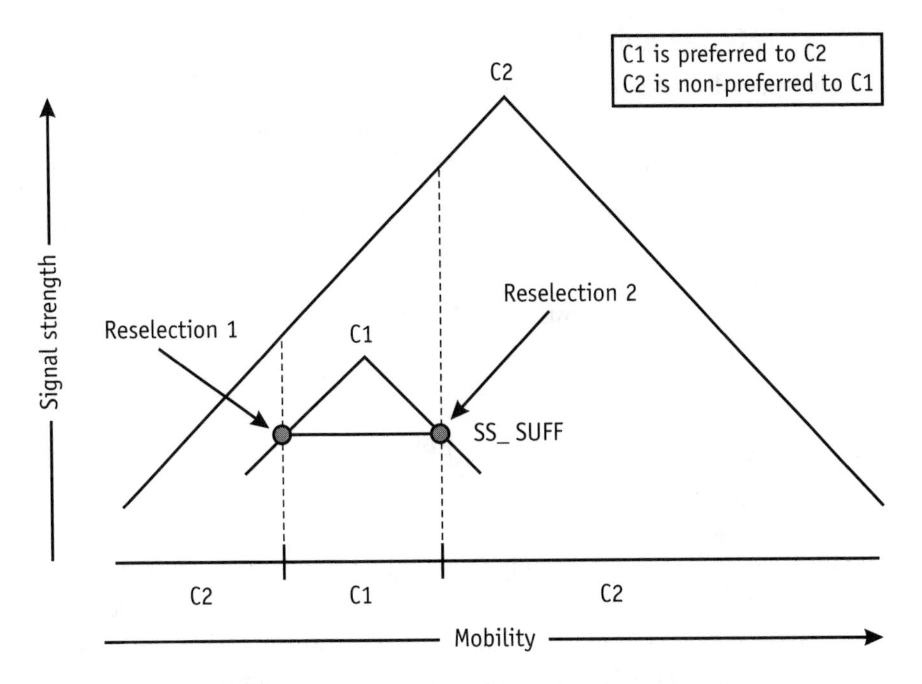

Figure D.3 Preferred and nonpreferred reselection scenario.

D.6.1 Regular neighbor

Figure D.2 shows a reselection to a regular neighbor. During this process, the phone will be evaluating the neighbor and applying the following algorithm (This is a simplified version—see "The Full Reselection Algorithms."). If the following reselection criteria, as well as the camping criteria, are met, the neighbor would become suitable for reselection.

$$\text{RSS [cand]} - \text{RSS [current]} + \text{RESEL\_OFFSET [cand]} > 0 \qquad \text{(D.3)}$$

If the signal strength of the candidate neighbor is greater than the current cell and a hysteresis offset (which can be positive or negative), the neighbor will then be suitable for reselection.

D.6.2 Preferred neighbor

An example of preferred reselection is shown in Figure D.3; it is indicated as *reselection 1*. Cell 1 is identified on the neighbor list as a preferred neighbor to cell 2. In addition to the camping criteria, one of the following criteria should also be met for cell 1 to be considered suitable for reselection.

$$\text{RSS [cand]} - \text{SS\_SUFF [cand]} > 0 \qquad \text{(D.4)}$$

or

$$\text{RSS [cand]} - \text{RSS [curr]} + \text{RESEL\_OFFSET [cand]} > 0 \qquad \text{(D.5)}$$

If the signal strength of the candidate neighbor cell 1 rises above SS_SUFF for cell 1, then that neighbor will be considered a suitable reselection candidate—or if the signal strength of the candidate cell 1 is greater than the current cell and a hysteresis offset (which can be positive or negative), then cell 1 will become a reselection candidate.

Note that SS_SUFF can be defined independently for each cell on a neighbor list.

D.6.3 Nonpreferred neighbor

Figure D.3 also shows an example of nonpreferred reselection between cell 1 and cell 2, which is indicated in Figure D.3 as reselection 2. In this case, the camping criteria and the following two algorithms must be satisfied.

$$\text{RSS [cand]} - \text{RSS [current]} + \text{RESEL\_OFFSET [cand]} > 0 \qquad \text{(D.6)}$$

and

$$\text{RSS [current]} - \text{SS\_SUFF [current]} < 0 \qquad \text{(D.7)}$$

The candidate signal strength must be greater than the serving cell signal strength and a hysteresis offset, in addition to the signal strength of the current cell dropping below the SS_SUFF of the current cell.

This case, which can be viewed as the reverse of the preferred reselection case, will occur when DCCH-capable phones leave a preferred cell (as in a private system).

D.7 DCCH reselection procedure

Control-channel reselection is comprised of three distinct and sequential procedures:

- *Reselection trigger condition* (RTC) procedure;
- *Candidate eligibility filtering* (CEF) procedure;
- *Candidate reselection rules* (CRR) procedure.

D.7.1 Reselection trigger conditions

The RTCs identify the conditions for which the phone will evaluate all the likely DCCH reselection candidates. The trigger conditions for control-channel reselection are the following.

- RTC1: Radio link failure;
- RTC2: Cell barred;
- RTC3: Insufficient DCCH serving signal strength;
- RTC4: Direct retry;
- RTC5: Service offering;
- RTC6: Periodic scanning.

The most likely reselection trigger will be periodic scanning RTC6 (which was explained in Section D.5), where the phone measures the neighbor channels to assess the DCCH environment.

D.7.2 Candidate eligibility filtering

The CEF procedure will execute the appropriate reselection algorithm described in Section D.6 for each neighbor on its neighbor list. In this way, all the different types of control channels can be judged in order to determine the reselection suitability in a multicell environment.

D.7.3 Candidate reselection rules

The CRR procedure uses the candidates identified by the eligibility algorithms to determine whether a control-channel reselection will take place. It also provides the criteria to determine the best control channel among the selected eligible candidates. This procedure enables all the regular, preferred, and nonpreferred channels to be ranked in order of reselection suitability so that the phone will camp on the desired control channel.

D.8 Advanced reselection—C_RESEL

The previous explanations assumed that the parameters RSS_ACC_MIN and MS_ACC_PWR were equal in the current and candidate cells. This will not always be the case, and the effect of a difference in these parameters is explained here.

IS-136 uses a variable called C_RESEL in the reselection calculations. This is not broadcast and is internal to the phones, but it is important to the way in which phones evaluate candidate neighbors—hence, the position of the reselection boundaries.

C_RESEL is defined as

$$\text{RSS} - \text{RSS\_ACC\_MIN} - \text{MAX}\,(\text{MS\_ACC\_PWR} - \text{P}, 0) \tag{D.8}$$

(The variables are as described in Sections D.2 and D.3.)

D.9 Regular reselection using C_RESEL

Using the definition for C_RESEL, the actual reselection algorithms for a regular neighbor relationship are the following.

$$\text{C\_RESEL [cand]} > 0 \tag{D.9}$$

and

$$\text{C\_RESEL [cand]} - \text{C\_RESEL [curr]} + \text{RESEL\_OFFSET [cand]} > 0 \tag{D.10}$$

Substituting from (D.8) results in:

$$\begin{gathered}\text{RSS [cand]} - \text{RSS\_ACC\_MIN [cand]} \\ - \text{MAX (MS\_ACC\_PWR [cand]} - \text{P, 0)} \\ - \text{RSS [curr]} + \text{RSS\_ACC\_MIN [curr]} \\ + \text{MAX (MS\_ACC\_PWR [curr]} - \text{P, 0)} \\ + \text{RESEL\_OFFSET} > 0\end{gathered} \tag{D.11}$$

Since RSS_ACC_MIN and MS_ACC_PWR were set equally in the previous examples, most of the terms cancelled out leaving

$$\text{RSS [cand]} - \text{RSS [curr]} + \text{RESEL\_OFFSET} > 0 \tag{D.12}$$

D.10 Different RSS_ACC_MIN between cells

If we assume that RSS_ACC_MIN is different in a current and a candidate cell but that the MS_ACC_PWR is set the same, (D.12) can be reduced to

$$\begin{gathered}\text{RSS [cand]} - \text{RSS\_ACC\_MIN [cand]} \\ \text{RSS [curr]} + \text{RSS\_ACC\_MIN [curr]} \\ + \text{RESEL\_OFFSET} > 0\end{gathered} \tag{D.13}$$

This can be expressed in words as:

> If the signal strength of the candidate cell is higher above its RSS_ACC_MIN than the signal strength of the current cell is above its RSS_ACC_MIN (plus the RESEL_OFFSET hysteresis), then the cell is a suitable reselection candidate.

This can also be expressed graphically, as shown in Figure D.4.

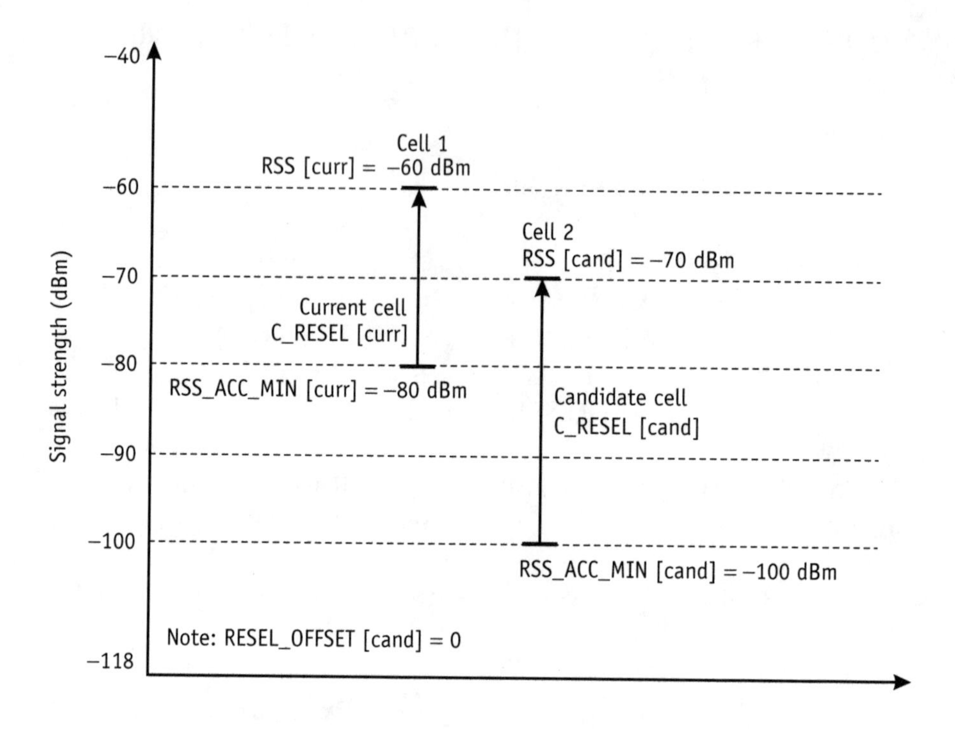

Figure D.4 Example of C_RESEL.

Using the example in Figure D.4,

$$\begin{aligned} \text{C\_RESEL [cand]} &= \text{RSS [cand]} - \text{RSS\_ACC\_MIN [cand]} \\ &= -70 - (-100) = 30 \end{aligned} \tag{D.14}$$

and

$$\begin{aligned} \text{C\_RESEL [curr]} &= \text{RSS [curr]} - \text{RSS\_ACC\_MIN [curr]} \\ &= -60 - (-80) = 20 \end{aligned} \tag{D.15}$$

Assuming RESEL_OFFSET is equal to zero, the criteria would be

$$\begin{aligned} \text{C\_RESEL [cand]} - \text{C\_RESEL [curr]} + \text{RESEL\_OFFSET} > 0 \\ = 30 - 20 - 0 = 10 \end{aligned} \tag{D.16}$$

which is greater than zero; hence, the candidate is a viable reselection candidate.

Note that in Figure D.4, although the signal strength of the candidate cell two was less than the signal strength of the current cell one, the 20 dBm difference in RSS_ACC_MIN for the current and candidate caused cell two to become a suitable reselection candidate.

D.11 Effect of difference in MS_ACC_PWR

Equation (D.11) showed that when (MAX MS_ACC_PWR – P, 0) comes into play it has the effect of reducing C_RESEL. This would shrink the size of a cell and disfavor it during reselection. This term is nonzero when the phone is unable to transmit at the output power requested by the base station and would be used in designing a rural rather than urban system.

Therefore, a nonzero uplink power requirement (MS_ACC_PWR – P) would tend to shrink the cell size for low-powered equipment.

D.12 The full reselection algorithms

Table D.1 details the IS-136 reselection algorithms for each neighbor type.

Table D.1

Full Reselection Algorithms

C_RESEL =	RSS – RSS_ACC_MIN – MAX (MS_ACC_PWR – P,0)
Uplink Power Requirement =	(MS_ACC_PWR ≤ 4dBm AND Mobile_Station_Power_Class = 4) or (MS_ACC_PWR ≥ 8 dBm)
Regular	1. C_RESEL > 0 and 2. C_RESEL [cand] – C_RESEL [curr] + RESEL_OFFSET > 0

Table D.1 (continued)

Preferred	1. C_RESEL > 0 and 2. RSS [cand] – SS_SUFF [cand] > 0 or 3. C_RESEL [cand] – C_RESEL [curr] + RESEL_OFFSET > 0
Nonpreferred	1. C_RESEL > 0 and 2.RSS [curr] – SS_SUFF [curr] < 0 and 3. C_RESEL [cand] – C_RESEL [curr] + RESEL_OFFSET > 0

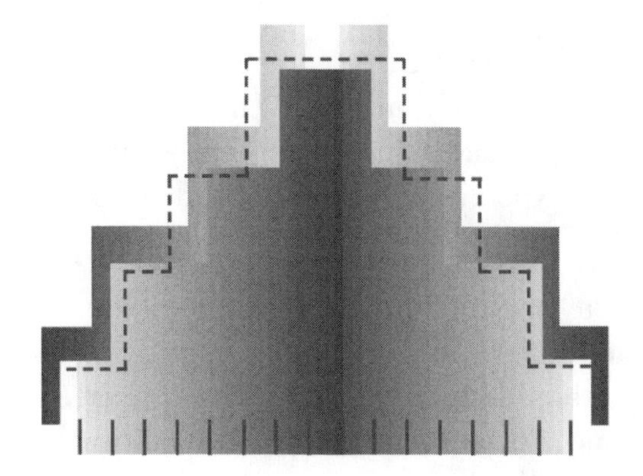

Glossary

A band The nonwireline radio frequency spectrum.

abbreviated burst A DCCH uplink burst length used when there is a large relative time offset from near and distant phones within a large cell.

AC *See* authentication center.

ACC *See* analog control channel.

ACC downlink message An analog control message sent from the base station to the phone.

ACC uplink message An ACC message sent from the phone to the base station.

access response channel (ARCH) A logical channel used for system response and administrative information. *See also* SMS messaging, paging, and access response channel (SPACH).

Advance mobile phone system (AMPS) An analog cellular telephony standard currently operating in the United States, Canada, Mexico, Caribbean, Central and South America, Australia, New Zealand, and part of Asia. AMPS phones may be used in other AMPS markets.

Algebraic code excited linear predictive (ACELP) A type of voice coding scheme used in the IS-641 EFR codec. This is similar to VSELP, in which pitch and other audio attributes are converted to code words for transmission, and provides improved voice quality performance under errored and normal radio conditions. *See also* vector sum excited linear predictive (VSELP), enhanced full rate (EFR), coder/decoder (CODEC).

Alpha tag A private system feature that places a system banner or company name on the display of a phone to tell subscribers that they have entered their private system. The banner is removed from the display when the user leaves the private system coverage.

AMPS *See* advanced mobile phone system.

analog control channel (ACC) An analog channel used for the transmission of control signaling information from a base station to a cellular phone, or from a cellular phone to a base station.

analog voice channel (AVC) A 30-kHz channel in the cellular frequency spectrum used to transfer FM speech information.

ARCH *See* access response channel.

ARQ *See* automatic retransmission request.

ATC *See* autotune combiner.

authentication A process used to stop cellular phone fraud. It uses a mathematically generated number, as well as the MIN and ESN, to verify a cellular phone for service. The authentication process in the DCCH systems use messaging on the control traffic channels to convey authentication information.

authentication center A network element that generates the A-key for the authentication process.

automatic repeat request (ARQ) An acknowledgment process whereby the sending element can retransmit blocks of data that were received incorrectly at the receiving element.

autonomous cells (system) Cells that broadcast a DCCH in the same geographical area to other DCCH systems but that are not listed as a neighbor on the neighbor list of the public system.

autotune combiner (ATC) An RF component that is able to automatically retune to different radio frequencies and combine multiple signals.

AVC *See* analog voice channel.

base station The radios and associated equipment at a cell site that communicate with a cellular phone in one direction and the switch (MSC) in the other direction. *See also* cell, cell site, mobile switching center.

Base station manufacturer code (BSMC) An IS-136 system identity structure that enables a phone to recognize base stations supplied by a specific manufacturer.

B band The wireline (phone company) radio frequency spectrum.

BCCH *See* broadcast channel.

BER Bit error rate. *See also* FER.

BRI field *See* busy-reserved-idle field.

broadcast channel (BCCH) The BCCH is comprised of logical channels that are multiplexed onto the physical DCCH downlink channel as part of the superframe. The BCCH logical channels include the F-BCCH and E-BCCH. The BCCH information includes system identification, neighbor lists of other DCCHs for DCCH reselection, DCCH frame structure, and other system information. *See also* fast broadcast channel, extended broadcast channel.

Browser An application in a phone that enables the subscriber to interact with a server application in the network.

BSMC *See* base station manufacturer code.

busy-reserved-idle (BRI) field Fields in the SCF field used to signal the availability of the RACH channel. A phone will then be able to determine if it can begin an uplink transmission.

C/A Carrier to adjacent channel interference ratio.

C-RESEL IS-136 uses a variable called C_RESEL in the reselection calculations. This is not broadcast and is internal to the phones, but it is important to the way in which phones evaluate candidate neighbors—hence, the position of the reselection boundaries.

call release messages Messages on the analog voice and DCCHs that indicate call release and may contain DCCH pointer information.

calling number identification (CNI) An IS-54B feature that allows a digital customer to view the telephone number of the person who is calling. A supporting capability, *calling number identification restriction* (CNIR), allows users to inhibit the display of their cellular number when placing a call. CNI is retained in IS-136.

camping This is the *idle* state of a phone when the phone is monitoring a control channel but not in a call.

candidate eligibility filtering (CEF) A procedure that executes reselection algorithms for each neighbor on its neighbor list. In this way, all the different types of control channels can be judged in order to determine the reselection suitability in a multicell environment.

candidate reselection rules (CRR) A procedure that uses the candidates identified by the eligibility algorithms to determine whether a control-channel reselection will take place. It also provides the criteria to determine the most suitable control channel among the selected eligible candidates.

CDL *See* coded DCCH locator.

CDVCC *See* coded digital verification color code.

CEF *See* candidate eligibility filtering.

cell Within a wireless system, the geographic area served by one or more radios at a cell site. Cells, depending upon design, can range in size from yards to miles. The traffic from a phone is handed off from cell to cell as the phone moves out of one cell into another. A cell might consist of up to three sectors, each serving part of the total cell coverage. *See also* base station, cell site, handoff, microcell, macrocell, sector, sectorization.

cell site The physical location, facilities, and equipment including the radio base station, antennas, and tower through which radio links to phones are established. The link between a cell site and a phone is a radio link; the link between a cell site and the switch (MSC) is landline or microwave. *See also* base station, cell, MSC.

cell type A parameter identifying the preference type of a neighbor cell. The cell type is set as either regular, preferred, or nonpreferred. It is used by the phone to determine which reselection algorithm will be used in the control-channel reselection process.

C/I Carrier to interference ratio (dB).

CMT Cellular messaging teleservice; a DCCH feature for transferring alphanumeric messages of up to 239 characters between the cellular system and DCCH-capable phones with a variety of attributes controlling the delivery, storage, and display behavior of the messages. The messages are sent and received via a message center. Also known as SMS, it uses the R-data teleservice transport mechanism on the DCCH or DTC to exchange the information.

C/N Carrier-to-noise ratio (dB)

CNI *See* calling number identification.

CODEC *See* coder/decoder.

coded digital locator (CDL) A DTC burst field used to inform the phone of the location of the DCCH.

coded digital verification color code (CDVCC) A DTC downlink burst field containing the eight-bit DVCC and four protection bits and used to indicate that the correct rather than cochannel data is being decoded.

coded superframe phase (CSFP) A DCCH downlink burst field that indicates to the phone which frame in the superframe is currently being transmitted. It helps the phone find the start of the superframe.

coder/decoder (CODEC) The CODEC converts voice signals from analog form to digital at the transmitting end and converts those signals back to analog at the receiving end. By digitizing the voice signal and compressing the resulting voice data, the bandwidth is reduced allowing system capacity to be increased. The CODEC can either be VSELP or ACELP. *See also* vector sum excited linear predictive coding (VSELP) and Algebraic Code Excited Linear Predictive (ACELP).

combiner *See* manual combiner, autotune combiner.

conversation state A phone is in this state when it is tuned to a voice or traffic channel and is providing a voice path for a call. The phone measures MAHO neighbors (if the phone is digital) and responds to hand-off commands from the system by retuning to the appropriate channel. When the conversation ends, the phone will enter the control channel scanning and locking state.

CRC *See* cyclic redundancy check.

CRR *See* candidate reselection rules.

CSFP *See* coded superframe phase.

CTIA Cellular Telecommunications Industry Association.

cyclic redundancy check (CRC) An error detection scheme used to protect information intended for transmission.

data link layer (layer 2) One of the IS-136 air interface layers. It handles the data packaging, error correction, and message transport. It provides framing and support for higher layer messages and attempts to ensure error-free transfer of layer-3 messages across the air interface.

DCCH *See* digital control channel.

DELAY A parameter defining the length of time for which a candidate control channel must meet the required signal-strength condition in the control-channel reselection process. It can be used to avoid the ping-pong effect at the cell boundaries.

deregistration A registration scheme through which a mobile notifies the system of its intent to leave its current network and reacquire service in a different type of network.

digital control channel (DCCH) A new control-channel mechanism described in the IS-136 specification that introduces new functionalities and supports enhanced features.

digital traffic channel (DTC) A portion of a 30-kHz channel in the cellular frequency spectrum used to carry TDMA digital voice information. Three such digital paths are generally available on each 30-KHZ channel.

digital verification color code (DVCC) The digital equivalent of the SAT, used to reduce the effect of decoding a cochannel interferer instead of the desired signal. *See* also supervisory auditory tone.

dropped calls Cellular telephone calls that are inadvertently dropped from the system because of interference, lack of capacity, or inadequate coverage.

DTC *See* digital traffic channel.

Dual-band phones Phones that can use both the 800-MHz and 1,900-MHz frequency band to gain the same IS-136 services.

dual-mode phones Phones that can be used on either analog voice channels or DTCs for wireless conversations.

DVCC *See* digital verification color code.

E-BCCH *See e*xtended broadcast channel.

enhanced full rate (EFR) The vocoder used in IS-641. This uses the ACELP coding scheme to provide improved speech quality. *See also* coder/decoder CODEC, algebraic code excited linear predictive (ACELP).

electronic serial number (ESN) The serial number of a wireless phone, which is automatically transmitted to the cell site every time a cellular or PCS call is placed.

ESN *See* electronic serial number.

extended broadcast channel (E-BCCH) A broadcast logical channel used for system information that is not required by a phone to gain access to a system but that will be required in later operations (for example, neighbor cell lists). *See also* broadcast channel (BCCH).

F-BCCH *See* fast broadcast channel

FACC *See* forward analog control channel.

FACCH *See* fast associated control channel.

fast associated control channel (FACCH) A blank-and-burst channel used for signaling message exchange between the base station and the phone.

fast broadcast channel (F-BCCH) A broadcast logical channel used for mandatory, time-critical system information requiring a fixed repetition cycle. The information sent on this channel relates to system identification and parameters needed by a cellular phone to gain access to a system in an expeditious manner. *See also* broadcast channel (BCCH).

FDTC *See* forward digital traffic channel.

FER Frame erasure rate. A more reliable measure of digital channel quality than BER.

forward analog control channel (FACC) An analog control channel used from a base station to a cellular phone.

forward digital traffic channel (FDTC) A digital channel from a base station to a phone used to transport user information and signaling. There are two separate control channels associated with the FDTC: FACCH and SACCH. *See also* fast associated control channel, slow associated control channel.

frame The largest unit in the TDMA timing structure. Each frame is 40 ms and contains six TDMA bursts.

frequency reuse The ability to use the same frequencies repeatedly within a system, made possible by assigning frequencies to cell sites in a way that no adjoining cell sites use the same frequencies. Frequency reuse allows a system to handle a large number of calls with a limited number of channels.

full-rate DCCH A DCCH system in which slots 1 and 4 are used to transfer DCCH information, in the same way that those two slots are used in an TDMA system to transmit one voice path.

general UDP teleservice (GUTS) A teleservice that uses the IS-136 R-Data transport mechanism to exchange UDP information. *See also* Teleservice.

GUTS *See* general UDP teleservice.

hand-off The process by which the system transfers or *hands off* a call to an adjacent cell. *See also* hierarchical cell structure.

HCS *See* hierarchical cell structure

hierarchical cell structure (HCS) An RF design structure introduced in IS-136 providing support for microcellular operation. A DCCH-capable phone will be able to reselect a particular control-channel neighbor cell over another based on the type of relationship defined between the serving cell and a neighbor cell. HCSs enable the DCCH to identify and designate neighboring cells as regular, preferred, or nonpreferred. The HCS designations are used by a phone to assess the most suitable control channel on which to provide service, even if the signal strength of a neighbor is not the highest being received by the phone, but is of a sufficient level to provide quality service.

HL_FREQ A parameter used when a phone has been in a static RF environment for a certain period of time, and the signal strength measurements for the neighbor cells have been fairly constant. Under these conditions, the phone is allowed to double the SCAN_INTERVAL to sample the neighbors at a slow rate. *See also* SCAN_INTERVAL.

HLR *See* home location register.

home location register (HLR) The network data entity that carries information about each subscriber and the registered phone's features.

hyperframe A DCCH burst structure made up of a primary and secondary superframe. The hyperframe length consists of 192 TDMA bursts—64 of these are used (every third) at full rate for DCCH information.

IMSI *See* international mobile station identity.

intelligent roaming The process a dual-band IS-136 phone uses to scan frequency bands and obtain the most suitable service. The 800-MHz and 1,900-MHz frequency bands can contain many service providers, some with incompatible systems. Intelligent roaming assists the phone in searching out the most suitable service as quickly as possible after power up. *See also* intellingent roaming database (IRDB).

intelligent roaming database (IRDB) The list of information downloaded to a phone using the OAP teleservice to support intelligent roaming. The database contains a list of frequency bands and preferred service providers that are used by the phone to search out the most suitable service as quickly as possible after power up. *See also* intelligent roaming.

IRDB *See* intelligent roaming database.

international mobile station identity (IMSI) An internationally unique number to facilitate seamless roaming in future global mobile networks.

IS-54 Interim standard issued by the EIA/TIA for cellular mobile telecommunications systems incorporating digital technology. It covers how digital radios and telephones work. The standard has gone through three revisions thus far, revision A, revision B, and the IS-136 specification (originally named IS-54C). *See also* IS-54B, IS-136.

IS-54B Interim standard issued by the EIA/TIA. Establishes technical requirements that form a compatibility standard for IS-54 cellular mobile telecommunications systems that incorporate TDMA technology. *See also* IS-54, IS-136.

IS-136 Interim standard issued by the EIA/TIA, originally called IS-54C. Establishes technical requirements that form a compatibility standard for 800- and 1,900-MHz mobile telecommunications systems that incorporate DCCH technology. The IS-136 specification is actually two documents containing the specification information necessary for development of IS-136-based products: IS-136.1 addresses the DCCH and the specifications for the DCCH air interface, and IS-136.2 addresses the modified air interface requirements for the ACC, the AVC, and the DTC. It consists of the existing IS-54B specifications with necessary enhancements. *See also* IS-54, IS-54B.

IS-54C *See* IS-136.

layer 1 *See* physical layer.

layer 2 *See* data link layer.

layer 3 *See* message layer.

logical channel A conceptual channel across the air interface. Many logical channels are multiplexed together to form the actual across-air bit stream. Examples of logical channels are PCH, F-BCCH, and SMSCH.

MACA *See* mobile-assisted channel allocation.

macrocell A cell site providing coverage over a relatively large geographical area (radius 1–5 miles).

MAHO *See* mobile-assisted hand-off.

manual combiner A passive RF component that combines RF signals from multiple sources.

mapping The method by which information is positioned onto a DCCH channel layer by layer until a TDMA burst is created, ready for transmission. At the receiving end, information is stripped off as needed.

MCC *See* mobile country code.

mean output power The calorimetric power measured during the active part of transmission.

message center (MC) A node on the wireless network to accommodate information sent and received via teleservices.

message layer (layer 3) One of the IS-136 air interface layers. It creates and handles messages sent and received across the air. Layer-3 messages are put into layer-2 packets that indicate the type of layer-3 information, the message length, the cellular phone to which the message is intended, and other administrative information.

message waiting indicator (MWI) An IS-54B feature available with dual-mode cellular phones that informs a cellular user of how many mail messages are waiting without the user having to call his or her voice-mail box. This feature is retained in IS-136.

microcell The hardware equipment and its associated antenna system that creates a relatively small coverage footprint. If the power amplifier of a macrocell is tuned to create a low ERP (10–500 mW/ch) at the antenna radiating point, the macrocell could be considered a microcell.

MIN *See* mobile identification number.

mobile-assisted channel allocation (MACA) A new function in IS-136 similar to MAHO, MACA is signal-strength reporting in idle mode on a DCCH. While on the DCCH, the phone measures signal strengths on specified frequencies and the signal quality of the current downlink DCCH. At system access, the phone includes a message containing these measurements. The base station, using the BCCH, sends a neighbor list informing the phone of where to look for potential cells for reselection. *See also* mobile-assisted handoff (MAHO).

mobile-assisted handoff (MAHO) A function in the digital system that allows signal-level (locating) measurements for the hand-off to be performed by the mobile. The mobile, in digital mode and under direction from a base station, measures signal quality of specified RF channels. The mobile scans channels on the downlink to be used in the hand-off algorithm. These measurements are forwarded to the base station upon request to assist in the handoff process. *See also* mobile-assisted channel allocation (MACA).

mobile country code (MCC) An identity structure that might be included in system broadcast information in support of international applications of IS-136 and international roaming.

mobile identification number (MIN) The telephone number of a cellular phone. It is automatically transmitted to the cell site every time a cellular call is placed. The MIN, along with the ESN, is used by the switch to identify the phone and its status.

mobile station identity (MSID) A 34-bit number enabling the current networking and authentication mechanisms to coexist with the DCCH.

mobile station (MS) Technically, a mobile phone attached to a vehicle transmitting with a standard 3W of power. As used in this application, however, MS and mobile station refer to both mobile and portable cellular or PCS phones unless specifically indicated otherwise.

mobile switching center (MSC) The facility that contains the cellular or PCS switch; it may also contain an HLR, service complex, and voice-mail system. The MSC processes cellular or PCS phone traffic in a service area. The MSC is the link through which a cellular or PCS phone user gets to the PSTN or another cellular or PCS user.

The term MSC is used to refer to the following.

- The physical location of the cellular or PCS switch;
- The equipment that does the call processing, which includes components in addition to the switch.

Other usages include the following.

- The MSC was formerly called the MTSO (mobile telephone switching office).
- The term MSC is often used interchangeably with the term switch.
- In the context of an SS7 network, an MSC is also known as a signaling point (SP).

MS *See* mobile station.

MSC *See* mobile switching center.

MSID *See* mobile station identity.

MS_ACC_PWR A parameter that defines the maximum nominal-output power that an IS-136 phone can use to initially access the network.

multiplexing The ability to send multiple signals over a single channel at the same time. Analog systems use frequency-division multiplexing. Digital systems use time-division multiplexing. *See also* MUX, TDMA.

MUX Multiplexer; a device that combines several communications channels so they can share a common circuit. Analog systems use frequency-division multiplexing. Digital systems use time-division multiplexing. *See also* multiplexing, TDMA.

MWI *See* message waiting indicator.

NACN *See* North American Cellular Network.

NAM *See* number assignment module.

NANP *See* North American Numbering Plan.

neighbor cell list Information for up to 24 neighbor cells that can be broadcast in each sector supporting a DCCH. Information includes channel number, cell parameters, cell hysteresis, and PSID indicator. A DCCH can broadcast information regarding analog control channel neighbors as well as DCCH neighbors.

nonpreferred neighbor cell In IS-136, hand-off and reselection to a nonpreferred neighbor will take place if the received signal strength of the serving cell drops below a certain threshold to provide service, and if the signal strength of the neighbor is greater than the current cell plus a hysteresis.

North American Cellular Network (NACN) An interconnection of regional cellular or PCS networks that allow cellular and PCS customers to travel anywhere within that network and still have their telephone behave and respond the same way as if they were in their home system.

number assignment module (NAM) The information specific to a cellular or PCS phone, such as its ESN, the phone number assigned to it, and the home system ID (SID).

OLC *See* overload control.

open systems interface (OSI) A seven-level standard communications protocol upon which SS7 is based. IS-136 is based on the lower tiers of this conceptual model.

OSI *See* open systems interface.

overhead message (OVHM) A message sent over the FACC that informs the phone of critical information, including data about system parameters, registration, and control channel information.

overload control (OLC) A means to restrict reverse control channel access by cellular phones. Phones are assigned one (or more) of sixteen control levels. Access is selectively restricted by a base station setting one or more OLC bits in the overload control global action message.

over-the-air activation teleservice (OATS) A teleservice that is used to automatically download the subscriber's information (such as phone number, SID, and PSIDs) into an IS-136 phone over the air, instead of through the keypad or by other means. This allows automatic programming of a phone and faster phone programming time for a customer. *See also* teleservice.

over-the-air programming (OAP) A teleservice that is used to program the IRDB into an IS-136 dual-band phone. This enables the phone to scan appropriate frequency bands and find the most suitable carrier and service as soon as possible after power up. *See also* intelligent roaming, intelligent roaming database (IRDB), teleservice.

OVHM *See overhead message.*

P An IS-136 parameter indicating maximum MS output power.

page The message sent out across the cellular or PCS system to find a phone that is being called.

paging The process of locating a phone prior to designating a voice channel during a network-initiated call setup.

paging channel (PCH) A logical channel used to transfer call setup pages to the cellular or PCS phone. *See also* SMS messaging, paging, and access response channel (SPACH).

paging frame class (PFC) The periodicity that an IS-136 phone will wake up to receive pages and other system information.

partial echo (PE) field An SCF field derived from a phone's MIN that echoes back part of the phone ID on the downlink. In this way, all phones will know which phone has use of the RACH.

PBX Private branch exchange. A landline telephone switching system for business use that connects internal stations and central office trunks to allow communications between system users, attendants, and the public network.

PCH channel *See* paging channel.

PCS *See* personal communications service.

PE field *See* partial echo field.

personal communications service (PCS) Sometimes used to refer to the 1,900-MHz band of operation, it is also used to describe the expected set of features and services from wireless technologies such as IS-136 (for example extended battery life, private system support, and teleservices). The DCCH is specified to operate in both the 800-MHz and 1,900-MHz frequency bands. IS-136 provides support for the same PCS features and services both during calls and while in idle mode regardless of the frequency band of operation.

PFC *See* paging frame class.

physical layer (layer 1) One of the IS-136 air interface layers. It deals with the radio interface, bursts, slots, frames, and superframes. The primary function of the physical layer is to map layer-2 information into TDMA slots, which comprise the fundamental transmission units sent over the wireless physical media.

ping-pong hand-off A situation that can occur when cells are marked as preferred to each other or when the cell parameters are set in a way that destabilizes the camping process—for example, when two cells in the same geographical area are considered the best candidates for each other, causing the phone to "bounce" between the two cells.

PL A parameter indicating current power level for a phone.

PREAM The preamble sequence in a DCCH uplink burst that, with the SYNC+ field, enables the base station to lock onto single bursts of data from phones and decode the uplink information.

preferred neighbor cell In IS-136, hand-off (generated by the system) or reselection (generated by the phone) will be made to a preferred neighbor even if the signal strength received from the neighbor is lower than the serving cell. The main criteria here is that the preferred neighbor cell must have signal strength defined by the system designer as sufficient to provide quality service.

private cells Cells that provide special services to a predefined group of private customers only and do not support public use of that cell.

private system identity (PSID) An identity structure assigned to a specific private system by the operator to identify that system to phones in the coverage area of that system. This allows a phone to determine whether it has special services from a particular cell when reselecting a DCCH. PSIDs can be assigned on a sector-by-sector basis, which allows very small service areas to be defined.

probability block One of 16 divisions of the 800- and 1,900-MHz spectrums, each containing approximately 26 cellular channels. IS-136 specifies the order in which a phone must scan these probability blocks when searching for a DCCH.

PSID *See* private system identity.

PSTN *See* Public Switched Telephone Network.

public cells Cells in the public cellular system that provide the same basic cellular service to all customers.

Public switched telephone network (PSTN) The telecommunications network traditionally encompassing local and long-distance carriers and now also including cellular carriers.

R/N field *See* received/not received field.

RACH *See* random access channel.

random access channel (RACH) The RACH is a shared bandwidth resource used by all DCCH-capable phones to access the system. The RACH is the only logical channel on the uplink of the DCCH.

received/not received (R/N) An SCF field used to indicate if the uplink message was received correctly. It is this field that provides the ARQ error-correction function on the uplink RACH.

RDTC *See* reverse digital traffic channel.

RECC *See* reverse analog control channel.

registration A function that gives a cellular or PCS system the ability to know the location and status of a phone. Existing AMPS and IS-54B registration schemes remain the same under IS-136, and several new forms of registration are introduced, providing for backward compatibility with existing systems as well as enhanced tracking of phone whereabouts.

regular neighbor cell In IS-136, a hand-off or reselection to a regular neighbor will occur if the received signal strength of a regular neighbor is greater than the current serving cell signal strength plus a hysteresis value, and there is no eligible preferred cell.

reselection In IS-136, a method by which phones reassess the best control channel for service other than just on determination of signal strength. IS-136 reselection can occur between two DCCHs or from a DCH to an ACC.

reselection trigger A procedure that identifies the conditions for which the phone will evaluate all the likely DCCH reselection candidates.

RESEL_OFFSET A parameter used to increase or decrease the preference of a candidate neighbor cell being considered for cell reselection.

residential cells (system) Cells that provide special services to a predefined group of residential customers only and do not support public use of the cell. The PBS would be classed as a residential system.

residential system identity (RSID) An identity structure that identifies residential systems within the public cellular coverage. RSIDs can be used to create residential service areas or neighborhood residential systems by broadcasting an identifier that is recognized by phones as being "at home" and therefore receiving special services (billing and so on). An RSID can also be used in the PBS, which allows an IS-136 phone to be used like a cordless phone in conjunction with a residential base station.

reverse analog control channel (RECC) The analog control channel used from a cellular phone to a base station.

reverse analog voice channel (RVC) The analog voice channel used from a cellular phone to a base station.

reverse digital traffic channel (RDTC) A digital channel from a cellular phone to a base station used to transport user information and signaling. There are two separate control channels associated with the RDTC: the FACCH and the SACCH. *See also* FACCH, SACCH.

R/N *See* received/not received.

roam *See* roaming.

roaming A cellular or PCS subscriber using a wireless phone in a cellular or PCS system different from the subscriber's home system.

RSID *See* residential system identity.

RSS_ACC_MIN A parameter indicating the minimum control-channel signal strength (received at the phone) required for a phone to access a cell sector.

RVC *See* reverse analog voice channel.

RVCH *See* recording of voice channel handling.

SACCH *See* slow associated control channel.

SAT *See* supervisory auditory tone.

scanning A procedure by which a phone searches for a viable DCCH candidate to gain service. This involves scanning the history list of last-used channels, investigating the ACCs for DCCH pointers, or scanning the probability blocks to find a DCCH or a DTC with pointer information. If a likely candidate is found, the phone executes a series of checks and comparisons (selection) to verify that the candidate DCCH is acceptable for camping and providing service.

SCAN_INTERVAL A parameter defining the time intervals during which neighbor cell measurements are performed by the phone. It is a tool designed to prolong battery life. The measurements can be performed each time the phone wakes up to receive pages. *See also* HL_FREQ.

SCF *See* shared channel feedback.

SDCC *See* supplementary digital color code.

sector Part of the coverage of a cell. A cell can be split into two or three sectors, each being perceived as an individual cell to increase coverage and capacity. *See also* cell, sectorization.

sectorization Early cell sites had all available voice channels operating on antennas that provide a 360-degree coverage pattern. When capacity or interference requirements determine that a 360-degree pattern will no longer work, the voice channels are divided into smaller groups, and each group is assigned to antennas that cover portions of the 360-degree pattern. This procedure allows for higher capacity and better control of interference. *See also* cell, sector.

selection A procedure by which a phone, after scanning for and finding a viable DCCH candidate, evaluates the candidate channel by executing a series of checks and comparisons to verify that the candidate DCCH is acceptable for camping and providing service.

semiprivate cells (system) Cells that provide the same basic service to all customers and provide special services to a predefined group of private customers.

semiresidential cells (system) Cells that provide the same basic service to all customers and provide special services to a predefined group of residential customers.

SERV_SS A parameter (nonnegative offset value) that can be used as a bias in the *service offering* cell-reselection process and is invoked autonomously by DCCH capable phones.

shared channel feedback (SCF) A DCCH downlink burst field that is a collection of flags used as a method of control and acknowledgment of information sent from the phone to the base station. This field controls and acknowledges information sent across the uplink physical layer.

short messaging service (SMS) The original name for CMT but now used to refer to the teleservice transport in general for CMT, OATS, OAP, and GUTS.

S/I Signal-to-interference ratio (dB). *See also* C/I.

SID *See* system identity.

sleep mode A DCCH feature to extend phone standby time and enhance battery life. This feature provides extended periods of time in which the phone can power down some of its circuitry and *sleep* between paging opportunities, thus saving battery standby time.

slow associated control channel (SACCH) A DTC downlink burst field used for signaling message exchange between the base station and the phone. A fixed number of bits are allocated to the SACCH in each TDMA slot.

SMS *See* short messaging service.

SMS channel (SMSCH) A logical channel used to support SMS teleservices including CMT, OATS, OAP, and GUTS. *See also* SMS messaging, paging, and access response channel (SPACH).

SMS messaging, paging, and access response channel (SPACH) The SPACH channel is comprised of the following logical channels: the SMS channel (SMSCH), which is used to support teleservices, the PCH, which is used to transfer call setup pages to the cellular phone; and the ARCH, which is used for system response and administrative information.

SMSCH channel *See* SMS channel.

S/N Signal-to-noise ratio (dB). The power relationship between a communications signal and unwanted disturbances within the transmission bandwidth. Analogous to a radio broadcast signal compared to background static or interference.

SOC *See* system operator code.

SPACH channel *See* SMS messaging, paging, and access response channel.

speech frame A block of compressed speech data from the CODEC. Each voice channel puts out two 20 ms blocks of data, or speech frames, per TDMA frame. *See also* frame.

SS_SUFF An IS-136 parameter that defines the minimum signal strength considered sufficient for a control channel to be considered for control-channel reselection.

superframe A DCCH burst structure made up of sixteen 40-ms TDMA frames equivalent to 32 consecutive TDMA blocks at full rate, creating a sequence of 32 DCCH carrying bursts spread through 96 TDMA bursts. Each DCCH burst in the superframe is designated for either broadcast, paging, SMS messaging, or access response information. The superframe structure is continuously repeated on the DCCH channel.

supervisory audio tone (SAT) One of three tones in the 6-kHz region that are transmitted on an analog voice channel between the mobile and base station to verify reliable transmissions.

supplementary digital color code (SDCC1, SDCC2) Additional bits assigned to increase the number of color codes from four to 64, transmitted on the forward analog control channel.

switch *See* mobile switching center.

symbol In the DQPSK modulation scheme, each symbol carries two bits of information.

SYNC *See* synchronization.

SYNC+ An additional synchronization field in a DCCH uplink burst that, with the PREAM field, enables the base station to lock onto single bursts of data from phones and decode the uplink information.

synchronization (SYNC) A bit pattern field in a DTC downlink burst used by the base station to synchronize the incoming data bursts on a TDMA DTC.

system identity (SID) The SID represents an international identification and a system number identifying the service area.

system operator code (SOC) An IS-136 system identity structure that enables a phone to recognize base stations belonging to a certain cellular operator.

TDMA *See* time division multiple access.

teleservice An application that runs over the air interface that uses the R-Data transport mechanism to exchange short amounts of higher layer data. Examples of teleservices are the CMT and GUTS.

temporary mobile station identity (TMSI) A 20- or 24-bit number representing a temporary mobile identity. It is assigned to a phone by the system at initial registration to provide enhanced paging capacity in the air interface.

TIA Telecommunications Industry Association.

time division multiple access (TDMA) A digital transmission scheme multiplexing multiple signals over a single channel. The current TDMA standard for cellular divides a single channel into six time slots, with each signal using two slots, providing a 3-to-1 gain in capacity. Each caller is assigned a specific pair of time slots for transmission at full rate. Half-rate operation enables six users to share one 30-kHz channel using one slot each.

TMSI *See* temporary mobile station identity.

traffic The communications carried by a system.

traffic channel That portion of the digital information transmitted between the base station and the cellular or PCS phone, or between the phone and the base station, that is dedicated to the transport of user and signaling information.

transceiver A radio transmitter and receiver combined in a single unit. A cellular or PCS phone uses a transceiver to send signals to, and receive them from, the cell site.

uplink RACH A DCCH uplink logical channel (phone to base station) used by the phone for system access.

upper application layers The IS-136 air interface layer that represents the teleservice currently being used, such as CMT, OATS, OAP, or GUTS.

vector sum excited linear predictive coding (VSELP) The algorithm used to code and decode speech information in IS-54B and IS-136 environments. *See also* coder/decoder (CODEC).

VMAC *See* voice mobile attenuation code.

VMLA Virtual mobile location area.

voice mobile attenuation code (VMAC) A three-bit field in the extended address word commanding the initial cellular-phone power level when assigning a cellular phone to an analog voice or traffic channel.

VSELP *See* vector sum excited linear predictive coding.

wireline cellular carriers Conventional wireline telephone companies also providing cellular phone service. *See also* B Band.

word Bit streams of binary-coded-hexadecimal code sent on the air interface to achieve analog signaling.

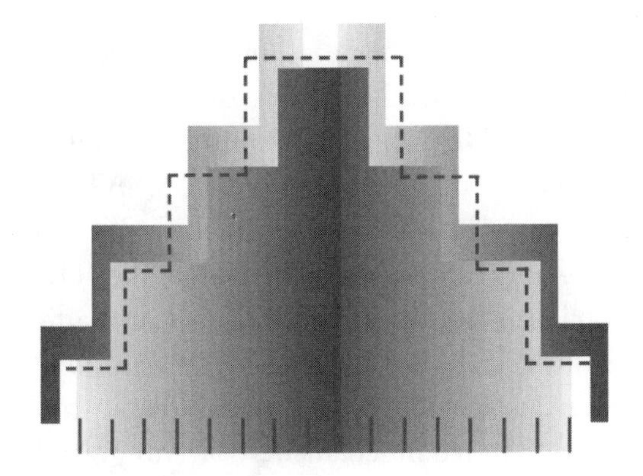

About the Authors

Lawrence Harte is the president of APDG, a provider of expert information to the telecommunications market. Mr. Harte has over 19 years of experience in the electronics industry including company leadership, product management, development, marketing, design, and testing of telecommunications (cellular), radar, and microwave systems. He has been issued patents relating to cellular technology and authored over 75 articles on related subjects. Mr. Harte earned his B.S.E.T from the University of the State of New York and his executive M.B.A. from Wake Forest University. During the IS-54 TDMA cellular standard development, Mr. Harte served as an editor for the Telecommunications Industries Association (TIA) TR45.3, the digital cellular standards committee.

Adrian Smith obtained a Bachelor of Engineering (Honours) degree in electronic and information engineering from Robert Gordon's Institute of Technology in Aberdeen, Scotland, in 1989. Subsequently, he spent 3 years at British Telecom Research Laboratories in the United Kingdom working on the ETSI standards for the digital European cordless telecommunication (DECT) technology; building DECT testbeds for voice, data, and video applications; and investigating DECT and ISDN interworking.

He moved to McCaw Cellular Communications (now AT&T Wireless Services) in 1993 and has since been involved in IS-136 TDMA—from the initial concept and development through the TIA standards process and onto field trials, phone testing, and system readiness for commercial launch by AT&T Wireless Services in 1996. Since then, he has been involved in the circuit-switched data, packet data, thin client architecture, and browser transport mechanisms for integrated voice and data solutions in IS-136. Adrian is the author of a number of patent applications associated with IS-136 technology.

Charles A. Jacobs holds a Bachelor's degree in Electrical Engineering from Georgia Institute of Technology. He is currently vice president of Product Management at Philips Consumer Communications LP. Mr. Jacobs worked for four years at Ericsson as product line director of TDMA Handsets, and prior to that, he was the business unit manager for Panasonic's in-building cellular system. He is a co-inventor of several intellectual properties in the wireless communications arena.

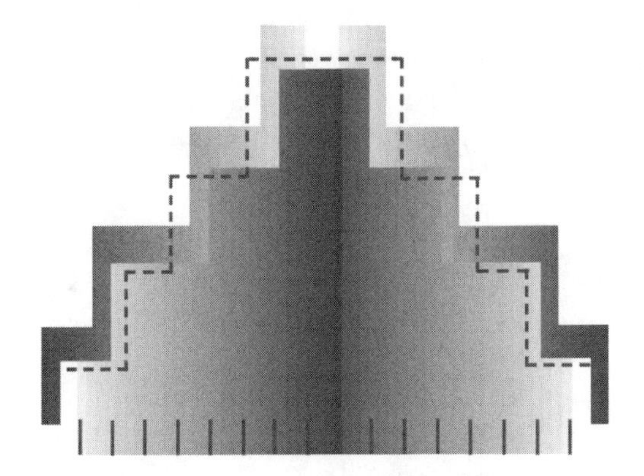

Index

The Artech House Mobile Communications Series

John Walker, Series Editor

Advanced Technology for Road Transport: IVHS and ATT, Ian Catling, editor

An Introduction to GSM, Siegmund M. Redl, Matthias K. Weber, Malcolm W. Oliphant

CDMA for Wireless Personal Communications, Ramjee Prasad

CDMA RF System Engineering, Samuel C. Yang

Cellular Communications: Worldwide Market Development, Garry A. Garrard

Cellular Digital Packet Data, Muthuthamby Sreetharan, Rajiv Kumar

Cellular Mobile Systems Engineering, Saleh Faruque

Cellular Radio: Analog and Digital Systems, Asha Mehrotra

Cellular Radio: Performance Engineering, Asha Mehrotra

Cellular Radio Systems, D. M. Balston, R. C. V. Macario, editors

Digital Beamforming in Wireless Communications, John Litva, Titus Kwok-Yeung Lo

GSM System Engineering, Asha Mehrotra

Handbook of Land-Mobile Radio System Coverage, Garry C. Hess

Introduction to Wireless Local Loop, William Webb

Introduction to Radio Propagation for Fixed and Mobile Communications, John Doble

IS-136 TDMA Technology, Economics, and Services, Lawrence Harte, Adrian Smith, Charles A. Jacobs

Land-Mobile Radio System Engineering, Garry C. Hess

Low Earth Orbital Satellites for Personal Communication Networks, Abbas Jamalipour

Mobile Antenna Systems Handbook, K. Fujimoto, J. R. James

Mobile Communications in the U.S. and Europe: Regulation, Technology, and Markets, Michael Paetsch

Mobile Data Communications Systems, Peter Wong, David Britland

Mobile Information Systems, John Walker, editor

Personal Communications Networks, Alan David Hadden

Personal Wireless Communication With DECT and PWT, John Phillips, Gerard Mac Namee

RF and Microwave Circuit Design for Wireless Communications, Lawrence E. Larson, editor

Smart Highways, Smart Cars, Richard Whelan

Spread Spectrum CDMA Systems for Wireless Communications, Savo G. Glisic, Branka Vucetic

Transport in Europe, Christian Gerondeau

Understanding Cellular Radio, William Webb

Understanding GPS: Principles and Applications, Elliott D. Kaplan, editor

Vehicle Location and Navigation Systems, Yilin Zhao

Wideband CDMA for Third Generation Mobile Communications, Tero Ojanperä, Ramjee Prasad

Wireless Communications for Intelligent Transportation Systems, Scott D. Elliott, Daniel J. Dailey

Wireless Communications in Developing Countries: Cellular and Satellite Systems, Rachael E. Schwartz

Wireless Data Networking, Nathan J. Muller

Wireless: The Revolution in Personal Telecommunications,
Ira Brodsky

For further information on these and other Artech House titles, including previously considered out-of-print books now available through our In-Print-Forever™ (IPF™) program, contact:

Artech House
685 Canton Street
Norwood, MA 02062
781-769-9750
Fax: 781-769-6334
Telex: 951-659
e-mail: artech@artech-house.com

Artech House
Portland House, Stag Place
London SW1E 5XA England
+44 (0) 171-973-8077
Fax: +44 (0) 171-630-0166
Telex: 951-659
e-mail: artech-uk@artech.house.com

Find us on the World Wide Web at: www.artech-house.com